Growing Goats and Girls

Growing Goats and Girls

Living the Good Life on a Cornish Farm

ROSANNE HODIN

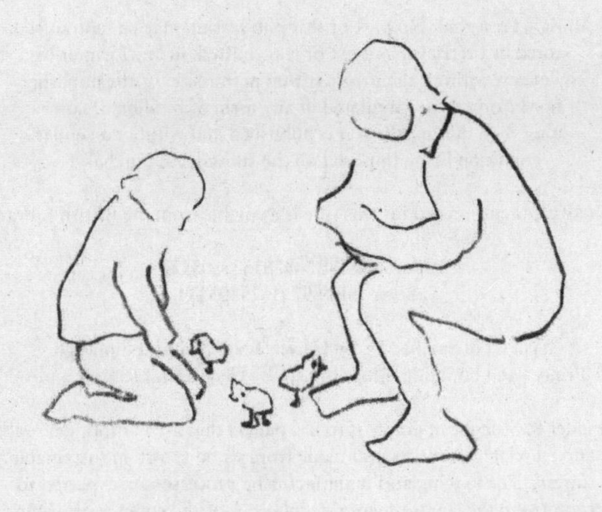

CORONET

First published in Great Britain in 202 by Coronet
An Imprint of Hodder & Stoughton
An Hachette UK company

This paperback edition published in 2021

1

Copyright © Rosanne Hodin 2020

The right of Rosanne Hodin to be identified as the Author
of the Work has been asserted by her in accordance with
the Copyright, Designs and Patents Act 1988.

All rights reserved. No part of this publication may be reproduced,
stored in a retrieval system, or transmitted, in any form or by
any means without the prior written permission of the publisher,
nor be otherwise circulated in any form of binding or cover
other than that in which it is published and without a similar
condition being imposed on the subsequent purchaser.

A CIP catalogue record for this title is available from the British Library

Paperback ISBN 9781529303322
eBook ISBN 9781529303339

Typeset in Sabon MT by Hewer Text UK Ltd, Edinburgh
Printed and bound in Great Britain by Clays Ltd, Elcograf S.p.A.

Hodder & Stoughton policy is to use papers that are natural, renewable
and recyclable products and made from wood grown in sustainable
forests. The logging and manufacturing processes are expected to
conform to the environmental regulations of the country of origin.

Hodder & Stoughton Ltd
Carmelite House
50 Victoria Embankment
London EC4Y 0DZ

www.hodder.co.uk

For Michael, Morwenna, Georgiana

Prologue

I am hunched and grey, spat out by the hospital. They have battled with my bleeding and doped and stupefied me and the baby has gone. So now I am home. Michael has gone off to the workshop stating that he has things to do. Being left alone to convalesce and deal with unwelcome hormones is being the star in my own bad movie. Outside, our neighbour Maurice is clanking in the yard with too many bits of metal and his Wolf Grinderette power tool. Obviously, I have to get up and take myself off before I leap over the wall and put the whining Wolf Grinderette through one of his arteries.

I am a bit dizzy and feeble, but I get the car out of the awkward parking space we have created out of wood chippings under the ancient beech trees and rhododendron jungle and I lurch into Liskeard. Nothing to do here, but buy cattle in the market, bank the day's takings, buy a pasty or gawp in the estate agents' windows. And there in the window is Large Bottom Farm, up for auction. I wobble in. The receptionist sees me eyeing the photo and says, 'Go and look. It's empty. Just drive there and let yourself in and look.' She hands me a key so heavy I nearly drop it on my foot.

Suddenly it is a perfect morning, lucid blue, thin crisp sky. I jostle down the country road, hovering in small wide places to let tractors pass. Go on beyond the hairpin bend and turn right. I get out to open the level-crossing railway gates, drive through and then shut them.

I have now cut myself off entirely from the world and as I drive up the last hilly metres to Large Bottom Farm, I feel a growing excitement. I stop and switch off the engine and sit and take in the silence and the trees and the river. No Maurice, no backyard car repairs, no walking or driving to the fields where we keep the goats . . . this could be just us and a sort of freedom. I put my hand on the dilapidated oak gate and step down a run of granite steps with the massive key clutched tight.

I hardly dare to go in. The house will be dank and musty and I will slink away and never even tell Michael of my moment of soaring hope. But the departing tenants of this farm have left it immaculate. Well, orange emulsion upstairs and fifties wallpaper with swirling silver oblongs downstairs, but clean and dry and smelling of welcome to my emotional nostrils.

It is a classic Cornish farmhouse, with a parlour, a sitting room, a large dairy with slate shelves, meat hooks on the ceiling and a pig-salting cask and an impossible kitchen with a decrepit oven range, a large open fireplace and a wormy corner cupboard. No kitchen sink, just an outhouse with a tap. Oh, and a bathroom with a huge, white enamel bath. I rush from room to room with a wild heartbeat. Please. Please.

I tell him. I drive to the workshop and order Michael into the car to come on a surprise trip. He humours me. I pull him round the barns, make him look over the fields, show him the house and then he faces me and says, 'You must be joking. I have just planted two rows of apple trees and a plum. Don't be ridiculous.'

Chapter 1

The auction at Webbs

The crowd inside Webb's Hotel is huge and consists of the most unlikely buyers of an empty farmhouse; most of all, it feels daunting. How could we have imagined that this could be for us, when there are hundreds of eager buyers with untold wealth in their pockets? We shuffle to the side, standing on the shabby carpet that covers the biggest of the Webb's reception rooms and we wait. We are auction house virgins.

The auctioneer arrives, looking the part in a flat cap and a tweed jacket with a checked shirt underneath. He tells us all what to expect of the procedure, the bidding, the need to have ready money, no backing out, and then starts up the process with such a low figure that the bidding comes fast with the auctioneer's song ringing and the nods or hands waving the property details. We stand dumbstruck and simply watch and wait. I hold Michael's hand tight.

As the figure rises, so the number of bidders decrease and suddenly it is our moment and Michael starts his own pantomime of a flick of the hand. Just two bidders now. Up and up. Everyone watches the silent race between us, racehorses in the last furlong. I'm writhing. Michael gives his last bid, over our agreed limit, and stops. So too the rival. The auctioneer takes a deep breath and says the auction is cancelled, the reserve is not met.

The tension crashes and the crowd is silent, before shuffling off with discontented murmurs.

The auctioneer asks us to stay. With a little persuasion, could we not raise our offer? The vendors would be so pleased to have us as the owners. The position of the land and the farmhouse is such that having owners like us would be to their liking. Us?

Er, um, well, yes? OK. OK.

And so, with the exchange of solicitors' details, we have a farmhouse, some land, an orchard or two, lots of barns, a river, trees, a privy, and a whole new future. And I am pregnant again.

Noah

So we move into a proper forever home with space for a family and for our goats. We move the goats into a hay-filled barn with a shelf to sleep on and a manger to hold hay, and better still a field, a river, space.

And so we go shopping for other creatures. We need a house cow. We need a Jersey house cow with dark eyes and a creamy tan hide and gentle manners and we have been told of a farmer with a Jersey herd and creatures to sell and so we go round to see. We are introduced to Rosetta. Farmer Hopkins tells us her age and breeding history and likely milk yield and how often she should be pregnant and then calve to keep up a steady milk supply. We are given her pedigree papers and herd number.

We will purchase a stainless-steel bucket that can be sterilised for the milking process. Mr Hopkins is astonished to hear that we intend to milk her by hand, but why would we want to set up electrical systems for one cow?

Rosetta takes her new ownership and solitary status with a placid acceptance and we persuade her to allow the hand-milking process. We lead her out of her shed by a rope halter and tie her to a metal bar not unlike a bicycle stand. She is given a bowl of oats and the milkman or milkmaid takes the old bathroom stool and hunkers down beside her udder and starts a rhythmic pull on her teats, which yield nothing until suddenly she gives in and lets down her milk with a great gushing flood, through our working fingers and into the bucket below, with a ringing sound as the jet of milk hits the side of the bucket. Creamy, frothy milk, full fat, almost yellow, enough to feed a huge family, more than enough for us; enough to make butter, yogurt, and skim off the cream.

We need pigs to eat the skimmed milk. Off we go to buy two weaners. We are learning pig talk. We transport our three beehives, consolingly covered with sacking. Then we buy Warren and Marran hens, point of lay, ready for a lifetime of egg laying. Ducks. Geese. Guinea fowl. Muscovy ducks. Soon our Large Bottom Farm is ringing with the sound of contented creatures. We feel like Noah and his missus.

Winter yard

It is bone-gnawingly cold. The animals have all tried to go to bed early, goats already in the stable looking balefully out of the door as if I am a bad waiter, endlessly and disappointingly late with an order. I swing the food barn door open to give oats and molasses-rich goat mix and see that the rats have been making merry in the feed bins. I will have to deal with them later.

The hens and ducks make a distracting shriek of enthusiasm and I gather up their assorted bowls and scoop out the right mix for each of them and I head out into the yard. There is mayhem under my feet as I open the gate. I have to extend my foot in a ridiculous ballet point to scoop away ducks from underfoot, plant my boot slowly amongst them and pick my way over to the duck house.

From the doorway I execute a spectacular move, well-practised and unerringly accurate. My right hand spins the corn into the duck house and as it scatters, a tide of ducks comes spilling over themselves and behaving like a mad rabble; my left hand shoots a spray of corn in the other direction along the concrete yardway to pacify the hens. This is the moment for separation of fowl: success.

Next I have to umpire an ugly scene between geese and hens and I shoo the geese into their shed. As soon as I feed the geese, their noise is simply a contented guggle of beak into straw as they pick up their grains. The hens busy themselves with the scatter in the yard. I must return to the rat issue.

I look around for something to plug the hole in the bin, but I can't stay out too long before I will turn into a pillar of ice. I pour a small but enticing pile of rat poison grains beside the feed bin and cover it with an upturned gutter section.

Quickly now, I fill the goat bowls and make my way unencumbered by fowl, as the hens, now properly suppered, have already gone in to roost. I pause in the goat shed to rub Plum at the base of her fat soft ears. She looks up at me, pleased, but has a more interesting job on hand with her oats. There is plenty of hay stacked in the manger still and I shut the doors, all creatures tucked away for the night and it isn't even 4 p.m. yet. I feel the pull of crumpets or muffins and a long slow evening ahead.

Snowdrops

There must be snowdrops somewhere. I am pulling on my wellies and hurtling out of the conservatory in a certainty that I will find a thin spear of white amongst all the mossy sprawl under the yew trees. I find crocus leaves beginning their untidy spikes under the apple trees, but I go on, desperate for snowdrops, desperate for that reassurance that the dark days are slipping away. Yes, over there, a small brave clump of six, seven, shivering slightly and gleaming. I crouch down and fill myself up with snowdrop.

Sawing horse

Cyril comes to look over our purchase; he knows about things. He walks the boundaries and gazes carefully at the barns and the fields. We trust his judgement more than anyone else's, knowing he left his family tradition of lightermen on the Thames and brought his red-haired wife to a tenanted farm in Wales to rear a family, to hand-milk a herd, to live without electricity, to work the land with sweat and blood. He is eighty, tall and strong. He looks at our land and finds it good. He offers wry advice here and there and comes back a week later with a gift of a sawing horse he has made for us.

It has two end pieces of diagonal crosses and long centre pieces and we can lay boughs and logs along this frame, which hold them as we cut them up. I have never seen such a thing before, but Michael has, and he is thrilled. We use it most weeks. We drag branches from the orchard, the riverbank, the thicket of laurel or the overbearing sycamores in the field boundary and make a production line. One of us heaves the branch to the sawing horse, one holds the branch steady, but it's always Michael on the saw or chainsaw, and one person has to nip about loading logs into a barrow. It's me.

The best thing is the hurling position. Lob a log high, higher, the highest, right to the back of the log store and hope it tugs itself down and doesn't start an avalanche. This is no pile for the orderly minded, no stacking in serried rows, it's a cheery wild mountain of logs, and we have borrowed a sign from the forestry workings in the wood, which announces: 'No Climbing on Wood Stack'.

Conservatory

Michael insists on a conservatory. It is an unknown around here. His great aunt Gwendoline had one in her splendid

Camborne town house, where she reigned as eccentric queen of a mining dynasty, and now we must have one for growing a vine and basking ourselves in sunshine. Two builders shake their heads, suck their teeth and slope away. Our friend Tristan is not so easily defeated and reminds us he has a failed degree in architecture; he rustles up drawings of a glass structure spanning the courtyard between the house and the washhouse and snouting outwards in a five-sided fan.

The main timbers are quickly assembled and then we have to cut away half: too many, too dark. Finding the reinforced triangular cut glass for the roof is complicated, but Tristan is resolute and while we wait for delivery, we tackle the enormous task of shifting the tomb-sized granite floor slabs that were once the kitchen floor into place as the conservatory floor. We dig hollow pits for them and then layer in sand and manhandle them into place. We stand back to admire the work in progress: planting beds on each side and a blank wall for a table and chairs.

Our heads are full of the thought of the vine, the ivies, the tomatoes, the seedlings, the hothouse plants, the children's sandpit, the hammock, and the lazy days ahead.

Planting a vine

Everyone knows that a conservatory should have a vine: we have a vine, a Black Hamburg in a little flowerpot. Everyone knows you should plant a vine on nutrients like a heart, a brain . . . a placenta? I have a placenta, I intend to use it for planting our vine, but in the hospital things get difficult. I have spent a great many hours giving birth to Morwenna in an unfriendly hospital with me shouting a lot and Michael telling me to breathe and then forceps and stitches, but a beautiful baby and then the afterbirth, push, push.

Well, it seems reasonable to me to ask the nurse who is trying to carry my placenta away if I can inspect it and she comes over with a look on her face that should warn me. 'Thank you,' I say and take the grey cardboard dish in which my nutrient-rich placenta lies; she is as quick as a flash and grabs it back.

'This is mine,' I say. 'I am going to take it home.'

'No, this is hospital property,' she tells me.

I point out it is mine, from my body, of my body, made by me, therefore my property and I am going to plant a vine on it.

A tug of war commences, where I pull on the cardboard dish and so does she.

'It's mine,' I say.

'It's against hospital regulations.'

'I want it.'

'You can't have it,' and she gives a stronger pull and scuttles off with it. I am furious and probably a bit deranged in a post-birth state. Michael is holding our newborn with a look of sublime joy, oblivious to the visceral battle at his side.

When we are home, he plants the vine on a chicken carcass. Harrumph.

The health visitor

At last there is a medical visitor who is not twitching with disapproval. Our house is in disarray and I have a baby. They don't like that. Before Morwenna was born, we were just moving in and the health visitor came with me to look at the farmhouse and was horrified that there was not yet running water. 'You can't move in here,' she said, 'you will have to be taken into Care.'

Care sounded worrying and so we squatted with kind friends until Michael deposited me, a crate of Guinness and a new baby with my parents.

But now we do have running water, and what does it matter if the kitchen is not finished? I sit and nurse the baby and give the builders cups of tea and we are all getting on with what we must do.

There is a new health visitor, who does not bristle. She tells me that babies are weaned with bananas in South America and that mothers do what they can. She thinks carefully about my worry that Morwenna cries and I'm not sure what to do and she tells me I have to check the key things: is she hungry? Is she wet? Is she the right temperature? If all those things are okay, then the baby has to toughen up, and I can wrap her in a blanket and push her pram to the end of the garden and let her shout or listen to the birds or sleep.

What liberation. Morwenna, you are going to the end of the garden. To sleep, please.

Yard work

It's a raw morning and there has been an exhausting baby-crying session, which seems to have stretched itself through the night. Michael pulls on his jeans, shirt, jumper, and steels himself out of the kitchen door. The boot beetle is in the way and clangs and skids across the granite floor. Fat socks. Wellies. Farm jacket.

The air outside is chill mist, the sort that claws your throat, and the gravel path down to the farmyard has collected small pools of wet. Primroses on the bank seem to issue a yellowish glow, unperturbed by the soaking of dew they carry. The whole bank is sodden, making the ferns, the dog's mercury, the campion all sag. Only the shiny discs of pennywort stand waxy firm.

Rosetta is the first in line for milking. Goats next. There are only three in milk and Yamaha needs milking only every other day. Michael takes Suzuki by the collar and they take a brisk turn down the yard. He reaches for one of the breakfasts and Suzy jumps up onto the milking shelf to eat and give herself up for milking. Michael pulls the white stool over and settles down.

It's a comforting business, head pushed to her warm flank, and she obligingly moves her leg back. One hand goes to grip the stainless bowl with a vice-like steadiness and the other pulls rhythmically on first one teat and then the other. It can take less than five minutes when all is going well and as long as the breakfast lasts. Once the food is finished, an irritable fidgeting starts up and puts the half-filled bowl in grave danger. Michael watches that frisky back foot as it starts to consider a little kick and he manages to pull the last jet of foaming milk and move away.

Suzuki is led away. Plum next, then the goat breakfasts are taken into the rest of the eager crew. Two bowls on the floor at opposite corners and one on the manger shelf. Everyone rushes to get to the best fullest bowl and then moves to see if another looks better or fuller and there is a fuss of bowl sharing and head tossing and fast movement around the shed.

The breakfast is short and quick and Michael leads Yamaha out into the meadow, where he tethers her to a running line. The others follow her lead and are easy to fix to their tethers. All that is left to do is to move one of the posts at the end of the running line for a new stretch of grass. He pulls up the post and clangs it in with the hammer. Four lines moved, four times the clanging sounds echo up and down the valley. I can hear them from my bed as I feed the baby.

Polly

Annoying noises from the farmyard are forcing me out of doors and I stumble into my boots, grab a jacket, and march out to see what mayhem is taking place in the animal kingdom.

A strange dog has a duck by the neck and is shaking it beyond death it seems, and all the geese and ducks are shrieking in protest. The goats have come in to see what the fuss is, nosey like me, but do nothing, unlike me, because I stride in through the gate and grab the black and white dog by its collar and wrench it hard. It looks at me in a mix of reproach and surprise and drops the dead duck. My temper flares and I hurl the dog into the nearest shed, the duck shed from which there is no escape, and I shut the door.

I check the duck in case there is life or a possible meal out of it and as there is neither I leave it on the lawn, for evidence. And I wait.

I go back into the house and do my waiting with a sleeping baby and a cup of tea. Soon enough I hear the cry of a likely owner, 'Polly, Polly, Pollleeee,' so I put my boots on again and saunter out into the drive. There is already body language at play. I imagine I am looking assertive and proprietorial; the man at the end of the drive is looking uncertain, supplicant.

'I've got Polly in a shed,' I call out and he comes nearer, looking relieved.

'I kept her there after she killed one of my ducks,' I continue in a neighbourly fashion.

I walk towards the yard and he follows, and I open the duck shed and there is a moment of owner/dog reunion.

'I kept her there so you could collect her and she could be reminded of the duck, and you can punish her. I daresay you will want to beat her when I show you the duck?' I'm saying this as a righteous challenge.

At this point the owner looks afraid. Clearly, he hadn't thought of a punishment being necessary. I have demanded it. I have the duck. It is still dead and has tell-tale bloodstains on it.

I offer a stick, but he chooses to wallop Polly with his hand, quite hard, shouting out in a fairly unconvincing tone: 'Bad dog, bad Polly!' It seems that Polly is as surprised by the beating as her owner, because she turns on him and bites him hard on the arm. He screeches and we both look at his bleeding arm.

'Hmm, you'd better come in for first aid. Polly can continue in the duck shed.'

This is one way of getting to know your neighbours.

Yellow sofa

We are ripping our farmhouse apart. It was politely arranged into four solid traditional rooms with a severance of a hallway and staircase in the middle. Each room has only one window. A parlour, a dairy, a sitting room and a kitchen. There was no downstairs water supply. I imagine the previous inhabitant walking out along a covered way to an outhouse for water heated with a copper tank in the corner. It's all change. Water and a sink in the kitchen, please.

The two rooms on the south side are knocked into one and the open fireplace is turned into a south-facing window and seat and the broken-down cooking range is gutted out and in its place is a brand-new Belling cooker, which is to last us thirty years.

Hey presto, we have light, sunshine pouring into the room. My dream kitchen is beginning. The next requirement is a table, chairs and a yellow sofa. The sale room has a table that we take to the acid-dipping tank at Menheniot station to strip it back to pine. Three chairs follow. The next sale has a pair of sofas, one an unwanted Chesterfield from the police station and the other a sturdy, old-fashioned piece, which might manage a young family. Then Amelia finds me the yellow fabric I need for a sofa cover.

A piano appears, hideous until painted white. A little table for tiny tea parties or playdough arrives. Yellow cotton curtains. The room morphs around its growing inhabitants, but the yellow sofa in all its different incarnations must stand the test of time, each collapse of springs necessitating another sale room visit with a tape measure to check that the yellow cover will fit. For thirty years this kitchen will be a place of sunshine and comfort and a yellow sofa.

Workshop in the sitting room

Flexible working conditions? We are the master of these. We have moved our glassblowing workshop into the sitting room. The room is large and predisposed to a division, so we have a sofa hunching its back across the middle of the room, separating a carpeted zone and a slate floored zone where Michael and Julie can sit and chat and work in warmth as they make filigree glass galleons, vases of mice running up stems of barley and other vital tourist items for our seaside tourist shop. They have worktables with a Bunsen burner, stacks of glass rods and the comfortable soft buzz of the gas flame.

The carpeted zone isn't getting much use. That is, until today, when we find we have a formal visit from the rector. The kitchen is a chaos of baby toys and drying towelling nappies, so I invite him for a polite cup of tea in the sitting room.

Here I perch on the edge of a chair with the baby, arrange him and his large black frock on the sofa and we exchange pleasantries. I am a newcomer; he is seeking out a potential member of his flock. He asks if I have planned a baptism and could he help? I stutter that I am thinking about dates and venues that will suit our unreliable chosen godparents, which is clearly the wrong thing to say, as he tenses up and says he has a few organised group baptism days ahead, and parents generally fit in with the rector.

I shift the baby and keep my face in a pleasant smile. The workforce behind me, with impassive faces, cynical views and scarcely held-in laughter, are straining to hear every word, to be used against me later.

Routine visit to the vet

I don't know which is worse: taking one's small children to be immunised knowing the hassle, the screeching and the maternal terror of presenting a perfect human specimen to be injected and punctured, or taking a soft new pair of twin goat kids to be castrated and have their budding horns burned away.

Today I must do the goat horror and know that a month ago I did the infant horror and I have been forgiven by her.

The goat logic is this: we do not want an immature goat rutting with his siblings or performing any of the malodorous male goat tricks on any member of our herd. And horns must be removed if we are allowing children to play with and amongst the goats. We then rear the young males to frolic and play and can eat them at nine months. It's farming.

The vet gives a local anaesthetic before he makes a swift incision and scoops the testicles out, then sprays the emptied sacs with a purple antiseptic. He anaesthetises the skull before he burns the growing ring around the base of the horn bud, with a dreadful smell of singed fur.

I take the kids home in a dopey state and their mother nuzzles them and leaves them to sleep in the hay. I worry about them for a while, empathise about their probable headache, but when I go out to the yard to check on them, they are bouncing about in the field apparently unaware of the violence we have had performed on them. All is well.

Priscilla and the garden

I've got a special guest arriving, to advise me on how to turn a sea of mud and random grassy patches and scraggy fruit trees into a garden. She is the wife of the bishop and is a botanist and a knowledgeable gardener. After coffee and when I have put the baby to sleep, we go out and stare at the problem.

Outside the kitchen door are a drab concrete passageway, outdoor washhouses, a tiny but bare courtyard, a cinder track, laurel bushes, bare conservatory, and a secret path to the outside privy. It's a muddle and indescribably ugly. I explain that what I want more than anything is to walk out of the kitchen and see flowers and loveliness. Priscilla can surely hear the desperation in my voice.

She walks to and fro, talking about south and west. I blurt out that I thought maybe I could pull up the laurel hedge and make a border for flowers and she nods, thinking.

She goes to her car and returns with some little hellebore plants and I scamper upstairs to get the baby, who is gurgling. I wear the baby on my hip and we carry on the survey of my dismal lawn.

She puts a hand on my arm and says, 'Why don't you just get some grass seed down and let that grow, and maybe pull up the cinder track and throw in a few tough and cheerful plants to keep you going? You have a baby; really a garden can wait, you will be so busy.'

Wise lovely words. Thank you.

Goat husbandry

We have a shelf devoted to self-sufficiency: *Goat Husbandry*, *Backyard Dairy Book*, *Backyard Poultry*, *Food for Free*, *Fruit Garden Displayed*, *Burcombes Queenies and Colloggetts*, *Plumbing in the Home*, *Wild Food for*

Goats, *Self Sufficiency*, and more; all are well-thumbed. Most have muddied fingering on the pages dealing with crop failure, pests, and surpluses.

Goat Husbandry intimidates us both. It speaks with a lofty expertise we cannot hope to match, with diagrams of custom-built goat pens instead of our crumbling stone barn, of careful fodder crops instead of our one meadow and old orchards and goats kept on tethers, of marketing a range of goat products in glossy containers instead of milk in glass jam jars. It has pages and pages of well-scanned lists of health problems and clearly every sickness or droopiness has had us reaching for the book to see what symptoms match.

It's a ghastly read.

Phosphorous deficiency. Poisons. Black Garget. Chronic Mastitis. Gas Gangrene. Foot Rot. Goat Pox. Pulpy Kidney Disease. Most of the cures listed involve diet change and many have the 'easy' tag: success or failure apparent by the ninth day. I imagine that failure is death with the irksome business of disposing of a body. One becomes resigned. Or one visits the vet.

The book has very helpful pages that show cross-sections of goat pregnancy, and the way a single or twin foetus can lie, and how one can extract a goat kid if necessary – 'bring that leg forward with your finger, if possible'.

There are diagrams too of Mendelian dominance and recessive genes with paragraphs on hornlessness hermaphrodites and risky matings. If one has a dreary rainy afternoon, what better read than an hour or two with *Goat Husbandry*?

Life crossroads

We have been playing for too long and need to grow up. Running a crazy, tinpot seasonal seaside business, where we aren't home till ten or later, is not working with our smiling fat baby, who has survived one season in a carrycot under the shop counter and breastfeeding that evoked a horrified response from our customers.

We have already had several careers apiece before departing from London for a life of self-sufficiency. We have been foolish, larking about when we should have been building a future. Now Michael is giddy with schemes, toying with mad ideas he immediately cancels until he comes up with The One. It's the career he has always wanted, he was just distracted into reading sciences at university: he wants to be a doctor. He's always wanted to be a doctor. He rushes full tilt into investigations; his old college will surely accept him again to read medicine.

The baby and I discuss this. Is she willing to be raised by only me for seven years while her father is a student far away? Do we sell the farm? End this new delicious life of self-sufficiency? Goodbye hens, goats, creatures great and small? Or do the baby and I run the whole show for seven years while Michael lives away?

I don't say any of this to my husband – I keep silent, holding my breath with dread. A person has to do what they want in life. It just seems at a woefully hopeless time. The baby and I watch progress.

The college has Michael's application and his marvellous persuasive letter and their admissions tutor rings when Morwenna and I are mixing slop for the pigs. When we throw off our wellies and coats, we can see Michael leaning against the kitchen door cradling the phone on his shoulder. He is making low noises of agreement and assent. He doesn't give me any signal of which way the conversation is going so I creep about making supper for Morwenna and getting cups of tea. Finally, the conversation ends and he joins us at the table.

Well? This is our life-changing moment.

'Well?'

'They have had a meeting and a long discussion,' he says. 'They think it would be a great plan in many respects. They would be happy to have me back, my previous degree might even mean I could do a year less, but they think . . .' His voice trails off and he looks out of the window. He gathers himself up and continues, 'They think I am probably too old; the training would be so long at this stage in my life, with a young family. They advise against it.'

He fiddles about with his mug of tea and I can see his huge disappointment.

At this moment, neither of us can know that he will become a teacher, a quirky, witty, inspiring teacher, and run one of the best science departments around, and elect to stay doing just that, oh so joyfully, without shifting about to take up senior posts or a headship and that we will stay at Large Bottom Farm for thirty years.

'He has just given us our life back,' I whisper to the baby.

Chapter 2

Bartering with Amelia

There is a hippy mood in this town. There is the cooperative café where childcare, cooking and serving are shared and taken in turns. We even have shared parenting of a kind. There is lots of open talk about conception, ovulation, breastfeeding. My friend Amelia has offered to take my small girl for a day a week. Morwenna is delivered full of breakfast and waves cheerily to me as I drive away. In a further wild fit of sympathy for exhausted mothers or a need to escape her own rural cage, Amelia suggests she can come and help me in the house and will accept barter as payment.

Today she has scrubbed and cleaned and polished and hung out the washing and we have had a glorious lunch together discussing proper grown-up things and we round off the lunch with coffee, a swap of red, fake-leather trousers and other items of glamour, and then the bartering.

Six eggs and a lettuce and some beans, she suggests. I explain that is ridiculous and pitiful if you link her labour to money and consider the cost of lettuce and eggs. She sighs. I need to forget money and think about the freshness, the purity. 'But lettuces are very easy to grow and the chickens just get on with the eggs,' I try to reason. She explains that the polishing and scrubbing was pretty easy. I offer a chicken, too. She accepts. We hug. We go off to catch a chicken, which is much more effort, even for women of the world.

Raspberries

This baby gets slung onto my back when I have work to do, or puts her face to my chest to snuggle down for a sleep, but today she wants to see the world and get excited and involved; she can sleep in her bed later while I pack things in little freezer bags.

We go out to the raspberry canes in the new orchard and balance a stack of trays and boxes on the stone hedge and I explain to her what fruit we are picking and what I will do with it. I hold a container in one hand and pick with the other. We have to bend quite often to find berries that have snuck under leaves and she squawks when the tipping up is too extreme. We work our way along first one row and then another.

I am reaching up and in and down and there are a few mouldy berries and lots more to ripen, still hard, little pale red or white knots, but the trays are filling up. I'm suddenly aware that the burble and chat from the baby has stopped and I have a sharp flash of worry that she has fallen asleep before she should. I put my fruit-stained hand on her head to turn it and look down. I find a beaming creature whose face is smeared with mushed berries, juice and the fragments of leaves. Her hands are red and sticky, her face is red and sticky, my T-shirt is red and sticky.

A cider orchard

One of the first exciting things about the farm was its orchard. To the left of the house, a hedge, behind the hedge an orchard of small apple trees; wizened, gnarly, old as old can be and laden with little fledgling apples.

No one was there to tell us about the apples, their variety or their age, so the obvious thing was to make a map of the orchard, marking each tree by linear position and naming it where possible: Short Trunk, Close to Gate,

Leaning Tree, and other imaginative titles. Then we would have to pick the apples carefully at the end of autumn and store them in serried ranks in the big barn and check them out during the late autumn, winter and even spring, to see how well they had kept and how they tasted.

The apple project went on being as exciting in action as it had been in planning. Oh, how tenderly we picked those little rosy things, laying them out so carefully, loving their sharp smell. We didn't face up to the fact that their taste was mostly sour; these might be a varietal that mellowed during storage.

The fortnightly check-up revealed a gradual spread of rot and shrinkage. These clearly were not keepers and even more clearly were not eaters. They were not even cookers, being so small and shrivelly. A bad crop. Maybe it was a bad year? Do apples sometimes have bad years?

I tell Jack-up-the-hill that the apples are disappointing, and he smirks the smirk that often appears when he knows something I don't.

'Cider farm, down there.'

So, they are ALL cider apples?

'They be.'

Damnation to all those wasted hours. We plan to pull them all up and begin again. But first, maybe we should make some cider?

A pig too fat

Our first pig is ready to take to the butcher. We are so proud of him; he is large and ugly and pale pink. We have rung around and learned that no one will kill a pig except at the big abbatoir in Launceston. Fine, we will take him there.

It's easy to lure him in – he follows the food bowl and nips up the wooden plank into the back of the van and Michael drives off at a brisk pace for only a very short while before worrying about the condition of the van. It is swerving weirdly. Inside the van, which is only a flimsy little thing, the pig is careering from side to side and Michael thinks the van might capsize. The road between Liskeard and Launceston is a winding, twisting and hilly one, so he takes the road less-chosen, via Pensilva, at a slow pace, incurring hootings from annoyed cars that can't overtake.

At the abbatoir, Michael asks anxiously how our pig can be sorted from the others, as ours is organic, home-grown and special. The bloke in charge snorts derisively. 'Your pig isn't cut, the meat will be tainted.'

And thus we learn about pig rearing:

Step 1: Castrate.

We learn step two from the butcher.

Step 2: Feed it properly (skimmed milk, pig nuts, rejected fish fingers). Our pig is just about useless, mostly fat with very little lean meat. Bulk for sausages only.

Step 3: Kill at its peak and not a moment later.

We are utterly chastened and put away those smug self-sufficiency thoughts we had been harbouring.

Morwenna advises on mowing

Morwenna is lashed onto Michael's back in the baby sling and her fat legs are thrashing with excitement and she is patting his head. He is taking her off on a mowing expedition. When they appear in the garden with the serious mower we bought in an auction – ex-council, green, solid, reliable and sporting a number plate – she gives me a happy wave and I wave back from the doorway.

The noise of the mower drowns everything else, but I can tell from the thrashing legs and arms that there is a fair old commentary going on and Michael's head twists back to answer her. As they pass me I blow a kiss, but Morwenna is far too occupied with the business in hand to see me and she is beating Michael's back with her small hands and shouting out, 'Faster, Pa, faster!' and with that instruction they hurtle down the slope out of view. By the ha-ha, I see them both roaring with laughter.

Muscovy ducks

No Muscovy duck shall ever again set fat, pink-webbed foot or feathery wing on our farm. The wretches. After all the feeding and housing they have been given.

I felt proud when we stocked up with creatures after arriving here: new life, new possibilities. New vegetable garden made out of a lump of harsh, caked field, broken down by double-digging and rotovating and manure and the sweat of Michael's brow and the aching muscles in his back.

That first spring there were rows of all the first bright things – spinach, lettuces, tiny seedling cabbages, their shy new leaves a miracle. We tended them so lovingly, with the baby lying on a rug beside us as we weeded. They grew inch by inch, watched over and watered.

We were sitting by the back door on the upturned slate pig-salting box when we heard the woosh of wings overhead and there were the three Muscovy ducks airborne. We had no idea those waddling uglies were creatures of flight, but there they were, transformed and soaring. We ran to the edge of the ha-ha and watched them floating in the air down the valley and felt a change of heart towards them, looking so lovely in flight.

Not so lovely now, though. As we enter the vegetable garden, we see three Muscovy ducks gorging themselves on our tiny seedlings and all that is left of our growing project are decimated rows of headless vegetables.

Tonight is the night of the sharp kitchen knives. Every Muscovy should tremble now.

Walking into the greengrocer's means that I have to endure the booming greeting from the bearded storekeeper, 'Hello, Goat Lady!' Of course, at this everyone in the shop gives up fingering the fruit and turns to stare at me. Do I have horns? Hooves? I have come to collect sacks of reject vegetables. The car is parked nearby and I seize the corners of the large paper sacks and drag or haul the heavy things along the pavement and lob them into the back of the car.

There are multiple journeys and I think the Beard Man enjoys his 'Hello, Goat Lady!' thing as I get it for each one. Maybe he doesn't want customers to think that he is selling dodgy stuff to dodgy people, me. But that is the hard part over and now comes the joyful part, when I can drive right down into the yard and call out to the goats that I have sacks for them.

The haul and drag from the car to the goat shed is easier and downhill and a festive gift, as the goats know all about it and have gathered to bleat and clamour. I struggle to get the sacks into the shed, because so many eager noses are poking into them and obstructing me. The choicest morsel is always a stalk of cauliflower and there are many in the sack, slashed off by the Beard Man, who likes to get rid of the eight or ten protective leaves that hide the curds. I get the leaves, the sacks are filled and the goats rejoice.

Beard Man's assistant doesn't understand the symbiotic relationship we have and is responsible for chucking in rotting pineapples or mauve tissue squares that have covered satsumas. The goats snort and discard these unlovely surprises with a toss of the head, keen to delve deep to find more stalks. Bruised apples are welcome. Yellowing broccoli, too. Stinking onions are not. So after the long supper of ecstasy, where all the goats jostle for a sack, I gather the smelly rejects.

By the time I am ready to leave the shed, the sacks are half-empty, and I roll down their tops to make the remaining contents more accessible. One goat has already got its head half-stuck, rearing up wearing the oversized brown paper bag as a hat.

What a fantastic boost to the goats' diet. What recycling. What a most satisfactory arrangement. The joys of the Goat Lady.

Toes for ducks

I have Morwenna on my hip, that small child who refuses to be put down or parted from me. I am attempting to feed the animals thus encumbered. It's mostly okay – amazing what one can do jostling the feed bowls, shifting the weight of this small girl as I shoot the bolt on the gate. The goats like this visitor, though, and tickle her fat, bare legs with their whiskery noses and make her laugh.

I scatter the corn for the ducks, which are circling us in a frenzy, and this causes laughter too. I am trickling the corn down slowly to make the enjoyment last and am disastrously giving the wrong message to the ducks, who stretch up and up to get to the grain and instead find something more interesting. Too late, I back away, but not before one duck has reached some toes and had a good tug at them, making Morwenna gurgle with pleasure until she realises her toes are being eaten and she howls. More in indignation than pain, I think, so we quickly nip out of the gate. She stands crossly and shouts, 'Naughty duck!'

Haymaking

Haymaking happens when it happens. Jack has arrived to bale up the long snakes of turned grass, which have been drying these last three days. It smells like heaven. I have wandered through the lines with Morwenna, whose chubby toddler legs sink deep into it and we sit and play.

As I bath my little creature, we can hear a Cornish drawl and the rumble and clatter of tractor and trailer down the drive.

'What's that?' she asks, and I tell her Jack has come with Colin to gather up the hay and squash it into bales for the goats and cows for the winter.

'Me, too,' she says.

'No,' I say, 'it's bedtime,' and I scoop her up in a tent of towel and get her ready for bed. She wants to look from her bedroom window and so I open it wide and we lean out to see the tractor and the fine cloud of dust that rises behind it. 'Me, too,' she tries again and I tuck her up in her cot and kiss her goodnight.

Michael and I heave the bales as they emerge from the machine and drag them to a central pile of sixes or sevens to load onto the trailer and then we run further along to repeat this until the field is studded with blocks waiting for loading. It is hot work and we are dressed in long sleeves to keep out the prickly blades of hay and I stop to flex my back and to wipe the sweat from my forehead and under my glasses. There is a small blob in the middle of the field and it's moving unsteadily towards us. I strain to see what it can be.

Morwenna. 'Me, too,' she says. This determined parcel has vaulted out of her cot, but how? Come down the stairs and escaped through shut doors, how? She has walked through the yard and through a gate over sharp stones with bare feet and is here smiling. We post her up into the tractor and bring in the bales.

Pig thoughts

I admire our pigs, but I cannot love them. They have unlovely faces. They have small, mean eyes and ugly mouths, although they do have exquisite little feet. They are clean and attend with diligence to their housekeeping, they mess in one corner, and turn their bedding of straw and fluff it up for sleeping in and under. I may be influenced by the rest of our animals, who hate them. Pigs are not considered fit companions in the field – the goats snort at them, and Rosetta stops and stares, so the pigs spend most of their days snuffling about on the concrete in the yard or rooting for interesting things in their shed.

Their appetite is frightening. Every morning we separate the surplus milk in the milking shed next to the pig sty, pouring the warm milk into the bowl of the separator and turn the handle faster and faster until the cream shoots out in a thin stream and falls into the waiting cream churn and the skimmed milk gushes into the pig bucket. The pigs, intelligent enough to know that these noises herald breakfast, squeal in anticipation. Their bran and oat mix is stirred in and we tip it into the trough and they slurp it down quicker than a dog eats his dinner.

They get a share of household scraps, too. The hens get some, but we offer chicken carcases and meat bones to the pigs and they love it. Pigs will actually eat anything.

I truly love the cow and goats, and I am just kind enough to the pigs and scratch behind ears or a long piggy back with sharp fingernails. This yields a low moan of pleasure. They wait for me at the gate to get their back scratching.

Yesterday our relationship fell apart. I came into the yard with an awkward bale of hay and they were clamouring at the gate. Balancing the bale while I swivelled to shut the gate, I slipped and fell, sprawled on my back in the mud. For the pigs everything had changed. I was no longer the kind purveyor of ear scratching, or the bringer of food, I was food itself. They snapped and rooted at me with open jaws for the few seconds I was on the ground, until I hauled myself up, and hanging onto the fence lashed out with a volley of kicks. I am boss around here, pigs, and I am not easy meat for you.

I left the bale on the yard floor and went back to the house to clean myself up and steady my shaken nerves.

My father and the cream separator

My father bumbles off down to the yard to watch the animals. He is trailed after by both cats, who are utterly devoted to him, but they leave him when he takes up position at the fence, leaning there to look at the pointless activities of our creatures. The cats don't consider this entertainment at all.

He joins in the morning feeding routine, carrying out bowls of oats, and he always makes deep chuckling sounds as he goes around the sheds, distributing largesse. He watches the milking, but doesn't take a turn. He is master of the cream separator. He pours the pail full of Jersey milk into the large zinc bowl, which is clicked in place on top of the container of twenty-seven stainless-steel cones through which the milk is spun. He loves the turning and spinning and his little churn of cream.

What he does not love is our system for washing. The big bowl is heavy and takes up the whole sink, but the container full of cones goes in the dishwasher, each cone prised apart, each sticky with cream, and the collection fills up the top layer. He huffs and puffs about all this. This is farmyard equipment, he complains, and should not be in the dishwasher along with our plates. I get on with storing the cream in the fridge, rolling my eyes at someone who might have time to wash up twenty-seven cones by hand.

Yew berries

Morwenna potters in and shows me what she has in her hand. There is a fistful of pale red sticky goo with black seeds at the centre, instantly recognisable as yew berries. I know these are poisonous and so does she.

'Oh,' I say, 'Noberries. You have been picking Noberries?'

'Yes,' she says, looking a bit crestfallen.

I keep calm and unpanicky. 'So what did you do with the Noberries?'

Morwenna, very slowly and carefully for her two years, says, 'I did eat two.'

'Hmm. So you ate two Noberries? But you aren't meant to eat any Noberries, remember? That is why they are called Noberries.'

She looks at me sadly.

Michele, my new friend, has come to visit and has witnessed this little conversation. I turn to her and say we must go to the local hospital to see what we should do. I am not sure how or in what way the poison works.

We arrive at the Passmore Edwards Hospital, where a keen nurse is in charge of an empty waiting room and we tell her that Morwenna says she has eaten two yew berries. She wants to know how reliable this small girl is in her counting, but I am unsure. The nurse moves away to her office and we are to make ourselves comfortable in the waiting room.

Morwenna finds a big box of toys and starts playing, while Michele and I look miserably at a poster opposite our slippery brown leather bench. The poster proclaims in six easy pictures the most common plant poisons. Yew is featured: eating yew berries can cause death from heart failure in a few minutes. This is not very cheering news. We look at the toddler in front of us poking the eye of a rocking horse and wonder how close she is to death.

The nurse returns. She thinks stomach pumping is too violent and extreme for a child of two – it would be better to give an emetic. I wonder how she will manage to persuade and then force any drugs at all down a child who is determined never ever to take medicine of any kind.

After a battle, the dose of emetic is in my child and she is clearly fighting off any urge to vomit. We sit and wait and wait as the rocking horse is rocked and then, at last, Morwenna gives a lurch and brings up two yew berries in a mess of spittle.

Digging potatoes

Morwenna has a tiny boiler suit on and is trailing after Michael as they go to dig up potatoes. He gathers up a long-handled fork and a trug from the shed and she carries the trug to the potato patch. It's early in the potato season and the leafy haulms are flowering. Michael plunges his fork into the soil below a strong-looking plant and levers it up and Morwenna squatting at his side darts her little fingers into the soil and brings up a gleaming potato the size of her fist.

Into the trug it goes and Michael loosens the earth again with his fork and she brings out another and another. 'How many?' asks Michael and she pauses to do a careful count. It's not enough for supper, so they start on another plant; he digs, she searches out the potatoes and brings them to the trug. A basket of pearls.

Rats and eggs

Dammit, we have rats stealing eggs again. I know there was a small clutch of eggs along the wooden ledge of the manger in the goat shed this morning and I left them there to encourage others that this was a good laying place, and now when I have returned, they are gone. Sometimes the hens turn cannibal and peck and eat their own eggs, but that always leaves the yolky messy shell. No, it must be rats. My poison trail is being ignored.

First stop is Cornwall Farmers to buy ceramic eggs, authentic-looking, encouraging to hens, unpeckable and surely obvious to rats. We buy four and place them on the manger ledge and in three other loved spots in the goat shed, fluffing the hay and making a cosy inviting nest. By mid-afternoon two hens are sitting and laying, we hope. I keep a keen ear alert to the cackling announcement of an egg and Michael zooms out to collect whatever has been laid before a rat has grabbed it.

He comes back in a mix of disbelief and rage. He was just in time to see a rat with an egg clutched to its belly and held on by a front paw. He yelled and roared and the rat dropped it and sulked off down a well-disguised hole in the hay. Dammit again. Archie cat, what are you doing? Where is your sense of responsibility? Go forth and be a deterrent.

Chasing a fox

I am in the food barn gathering oats and hear Michael shouting. 'It's a fox, a bloody fox has taken a duck!' I poke my head out to check what I think I have heard, 'What? A fox?' And Michael from across the yard shouts back, 'A fox has carried off a duck right under my nose and is off across the field.'

'Get it!' I shout back, but he replies, 'Na, it's far off by now.'

I'm furious, mad, and rush down to the gate and can see that the fox has paused in the next field. I'm off. Down past the dip that was once a pond, I hurl myself over the gate into Jack's field, throw back my head and start to roar like a lion. Jack's herd of red cattle look up at me astonished. They really like the roaring and as I run towards the fox, which has MY duck in its jaws, I see that they have started to run too, maybe sixteen of them. We are an impressive army of females.

I keep roaring and some of the cows join in with a bellowing noise and we are galloping after that fox. We are all girls against the enemy and feel powerful. The hooves on the ground are loud and the fox turns to look.

Seeing this fearsome sight at his heels, he makes a sound judgement and drops the duck and runs. Twenty paces, fifteen, ten and I get to the duck to find what I imagine will be carnage. But the creature is alive with a tiny, wild heartbeat, so I pick it up and tuck it under my arm.

I'm no longer roaring and my maddened army of cows have stopped in surprise to gaze at me and the duck and then they turn back to eat grass.

The duck and I pick our way back through the grass and over the gate and into the yard. Her little heart is still beating fast as I put her in the duck shed. When I return that evening, she isn't there. I whistle the ducks in with their night feed and count them. They are all present and I can't even tell which one was the captured prisoner.

Herd marking

The inspector from the Ministry has decreed that our creatures should have individual markers, by which he means bright yellow ear tags. He presents assorted leaflets that tell me about clamps like secateurs for fixing the ear tags, which I can bulk buy. I look at the leaflets gloomily and assure him that each goat will be individually marked when he next visits and a relevant mark for each will appear in the Animal Movement Book, which pleases him greatly.

There is no way I am going to blow the family's housekeeping money for the week on ear-tag machinery. When he leaves, I get out a permanent black felt-tip pen and go out into the field. The goats look up and crowd around me, curious to know the reason for my visit. I rub Yamaha's head and deftly draw three bold black blobs on her right ear. Then I put two blobs on Plum's right ear. The other goats don't like the look of this and back off, so I plan to blob-tag the others tomorrow. I march back home to record my tagging system of blobs in the book.

Yamaha, herd leader: marker on right ear: three blobs

Plum: marker on right ear: two blobs

The blobs available for tomorrow are: one blob, right ear; one, two and three blobs, left ear; but that identifies only six goats. My plan is already foundering. We have sixteen goats. Four blobs in a little square? Dashes? Five blobs? A bit like a dice? Dominoes? Hmm.

Herd book

As a keeper of animals, I need to keep this Animal Movement Book in which I record the arrival and departure of creatures. It's a scruffy, scribbled book with names of goats born (Yamaha, Suzuki, Mango, Baboushka, etc., etc.) and those slaughtered and eaten (the boys, Nike, Reebok, etc., etc.). I go with book in hand when I buy a pitiful mewling calf at market, since I like Rosetta's calf to have a friend to play with and share the milk. We also record poultry bought and sold.

At the market, each sale has to be signed for and at home we can be visited randomly by an inspector who needs to look at the book and sign it. I keep forgetting to make appropriate entries, then have a frenzied session of writing which creatures have gone for slaughter or been lent out, and finding the dates is almost impossible. According to the Ministry of Agriculture, Fisheries and Food, I also have a herd that is to be entered on the British Register of Brucellosis Accredited Herds.

My herd is Rosetta and calves. As part of this accreditation, Rosetta will be tested periodically for bovine ailments and the vet will appear and sink thankfully into a comfy chair in the kitchen and have a cup of tea. We are his easiest herd.

Oh Georgie

Oh Georgie, I am so sorry to report the sodden weeping of my pregnancy with you. I just cried and cried. I lay on the yellow sofa in the kitchen and stupid, unwarranted, unstoppable tears drizzled out of me and onto a pile of cushions. I was perfectly well, I was reasonably rested (given a toddler of two pattering about), but my hormones must have been to blame. You were a much-wanted pregnancy. You followed another miserable miscarriage. We were all excited, but those tears simply flowed.

I am lying on the sofa, huge. I expect the tears have just finished, or are about to begin again. The television is on with at least an hour of programmes for small people. I am drifting in and out of sleep. Morwenna is sitting cross-legged, joining in snatches of song. I surface out of sleep to hear her burbling away. When I next emerge from the sobbing, our kitchen has been transformed.

A checked blanket is spread out on the floor, its edges immaculate and neat. On top of the blanket lies a small blanket, equally neat and positioned exactly in the middle. On top of that, a tea towel. Around the edges of the checked blanket is a military line of flattened socks taken from the laundry basket. They have been smoothed and each toe and heel is pointing the same way, with the same precise distance between them. The line starts with the big black socks, pairs, single coloured ones of mine, little bright-patterned ones, starry ones, pink, green, spotty in diminishing sizes with their toes all pointing to the fireplace.

Morwenna herself, as I wake, is busy in the middle of the tea towel, this time lining up her entire family of Baby Williams. These are a remnant of useless shop stock, a carton of 144 matchboxes in each of which is crammed a plastic-headed, cloth-bodied bean baby. The Williams come in different colours and she has colour-sorted her entire family along the edge of the tea towel. Every Baby William is tucked up under the tea towel in a massive bedtime, with little arms sticking over the cover.

Morwenna is flushed and delighted. Her entire world is under control. She is in charge and I must lie here, too emotionally drained, too soggy to move.

Chapter 3

The rat on the doorstep

The midwife who has called to check up on me is clearly flustered. It's her second visit and this new baby is only three days old. I haven't been downstairs yet as Michael and his mother are in charge of all domestic arrangements. She is looking after Morwenna and cooking lunch and Michael is at school, but coming back as soon as the bell rings to sort out everything else. The new baby is doing a lot of feeding and crying.

'Is everything all right down below?' asks the midwife, and I'm not quite sure what sort of 'down below' she means. She sees the difficulty and clarifies, 'Downstairs, your mother?'

'No,' I reply, 'my mother-in-law, but why do you ask?' I sense there is some sort of problem.

She reddens a little and then says, very gently, 'It's the rat. Yesterday on the doormat, outside in the conservatory, by the kitchen door, well, there was a rather large rat and I stepped over it to ring the bell when I came to see you. Well, it's still there today, this rat, a large one, dead, and there may be a risk of infection with a new baby in the house. I wonder if someone might remove it?'

I laugh. In this frantic household right now a rat on the doorstep is a small irrelevance. Once she has gone, I put on slippers and a dressing gown and creep downstairs and pick up the rat in a fold of newspaper. I imagine that this rat is a gift of love for me, caught by Archie, but anyway it still goes in the bin.

Mr Warboy

Old Mr Warboy has an allotment at the edge of the railway line and as we pass we wave or chat. He has a house on the corner and a sagging mesh chair overlooking his garden and the world. I took a photo of him standing at the corner of the allotment with his baggy woollen shorts and a vest and a great armful of gladioli.

He must have noticed my advancing pregnancy and somehow heard that I had delivered a baby, because he is knocking on the door. In both hands is a strange and very heavy metal object, on his face a beatific smile. He asks me how the baby is and then looks down meaningfully at the object in his hand.

It looks like a large part of a train, but he tells me, 'It's a mangle, special for cutting turnip in the old days, for pig feed, like; you have pigs, see, and this mangle will come in good for cutting garden things for them pigs.' He offers it to me, proud and happy to be bringing such a useful gift.

I stagger under the weight of it, quickly bringing up my other hand to take the brunt. We both look at it silently for a while and then I thank him very much and tell him I will be out in the yard again soon enough, but right now the new baby is taking much of my time. He smiles and I smile, and a great, grim vision of mangling turnips and sprout stalks into pig feasts passes darkly over me until the squalling noise of a baby brings me back.

Ernest's daffodils

My mother-in-law has gone home, the baby is a week old, Morwenna is at nursery. I am sitting on the sofa in a trance when the jangly ship's bell announces a visitor. I am still sore, so I creep slowly over to the kitchen door and through the conservatory and let in our opposite neighbours, Ernest and May. Mr and Mrs McGregor, Morwenna calls them, because of the neat rows of vegetables and flowers visible across the valley from her bedroom window. In Ernest's hand is a moderate-sized bunch of daffodils of different colours and sizes.

'How pretty,' I say and reach out for a jam jar to put them in. I turn back to ask them about their day and to show them my small and utterly perfect baby. They seem to agree. I offer a cup of tea and May joins me at the sink. She puts her hand on my arm and murmurs, 'Dear, not just a pretty bunch of daffodils; you need to know that Ernest this morning went out and picked one bloom from each of his precious collection of museum daffodils; they are an honour just for you and the baby.'

I understand entirely and take Ernest his tea, and sitting at his side with the baby tucked in the crook of my arm, I say, 'Tell me about the daffodils. The little pink one is exquisite.'

He smiles and tells me . . .

Honey harvest

Transformation from bees and beehives to honey in a jar. It is one of the most exciting harvests imaginable, but it takes time and planning. We usually leave taking the wooden frames away from the hive until mid-summer, so the bees can bring in the limited heather nectar we have round here and we can also feed them end-of-season sugar as a top-up.

Lifting the heavy wooden boxes full of honey-heavy frames is step one and the bees are regularly annoyed by this piece of thieving and have to be pacified with puffs of smoke to distract them into thinking there might be a fire. We carry off the boxes into the kitchen, where the table is spread with newspaper and sheets of plastic.

On the table top, Michael has set up the enormous tin-plated drum with a handle, which is the centrifugal spinner to spin out the honey from the frames. It holds a great many frames at once and is a spinning job only for the strong and mighty. At the bottom of the drum is the tap for pouring off the honey once it has been spun out of the honeycomb. All this lies ahead; for now, we must take out the frames one by one and start the capping process.

I hold the frame wedged between hip and arm and start slicing the tops of the honey cells in a firm steady cut, as if it was a loaf of bread, and the knife slips easily into the honey or jags from time to time when it gets stuck in wax. The scrapings of wax, dripping with honey, are quickly thrown into a waiting bucket, but have already covered my hands and arms with stickiness. There is no point in wiping or washing, just keep going. Only one side needs cutting and the frame can be slotted into the waiting drum. Michael and I now have sticky hands and arms and our feet are making sticky noises on the floor.

It's hot in here, because the doors and windows are firmly shut to keep out inquisitive bees, who may well wish to take back this stolen honey.

We fill the drum and Michael cranks up the handle and it starts to turn, slowly at first and then faster and faster. The lid is shut, but we imagine thin threads of honey shooting out of the honeycomb and onto the sides and down to the bottom of the drum. We stop after a while and peer in. There is a pool of golden honey at the bottom.

I have a box of jars ready and take them one by one to the tap and let the clear honey pour in. Eighteen, twenty-four, thirty . . . the box is filling and the spinner is almost empty. So we return to the capping process, the slicing, the loading and the spinning, on into night time. The children are fast asleep in bed and we have to finish the job tonight.

We have a bumper honey harvest and when all the honey has been scraped from the inside of the spinner and the bucket of waxy cappings has had the honey drained off, it is time to clear up. We take the boxes and empty frames and the spinner out to the field, close by the hives, for final cleaning up by the bees themselves. The boxes of honey jars are lined up on the kitchen floor and I wash and wipe the table, the floor, the counter, every place where there has been honey.

On my knees mopping the floor, I see a small but unreachable splodge of honey under the fridge and think to myself, I will get to that in the morning . . .

Bee crisis

Aaargh. My laziness last night has plunged me into trouble. I take Morwenna to playschool and bring baby Georgie back for a day of summer and when I open the door the kitchen is full of bees. The window is open: a foraging bee has found the spilt honey under the fridge and has gone back to the hive to do her bee dance, informing the whole hive of a honey source in my kitchen.

What to do? Certainly get rid of the baby. I drive up the hill and thrust Georgie at Mary, Jack's wife, explaining that I have a bee crisis and will be back as soon as I can manage. Dash into the washhouse and put on a bee suit and beekeeper's hat. Sigh deeply as I enter the kitchen. It is alive with bees. Bees spread good news about a source of honey and carry on looking long after the source is finished. So the malingering bees have to go and no more are to come in.

With a heavy heart, I shut the window and door and begin a bee massacre with the potato masher. As a bee flies to the window to leave, and possibly to tell more bees that they should come in and look, I squash it against the glass. And again and again. I don't know how many bees I kill, but it is vile and I hate it. I am protecting my family, I tell myself, as I wipe away crushed bodies, clean off the windows and sweep the floor, and then shed my white boiler suit and hat. Murderer. Lazy fool.

Screaming babies

Our new baby will not stop crying. She snivels and wails as I hold her on my hip and jig her about and we all are in despair. In misery, I have watched her pinched, exhausted face and Morwenna and Michael have formed a gang dedicated to avoiding our company when they can. Our infinitely kind GP and his wife came to spend a day observing this squalling child and concluded that there was nothing medically wrong with her, but that maybe her early bouts of colic had taught her an ear-wrenching way of expressing misery, and now she probably wails like this out of bad temperedness.

Their diagnosis. A bad-tempered baby.

Morwenna gets a break from it at nursery, but we have to grit our teeth and cope with the afternoons. Then something wonderful happens at 3.30 p.m., when the school bell rings to mark the end of the school day; Michael simply leaves his desk, his marking, his responsibilities and he drives home at full speed and bursts into the house to rescue us all. He grabs howling Georgie and dances about with a delighted Morwenna and lets me throw on my boots and run outside into the freedom of the fields.

I run to the river and splash in the shallows, I stroke a goat's nuzzling head, I pick supper vegetables, I lie in the grass and stare at the clouds, I gather up the feed bowls for the animals and fill them with oats and other mix, I stand on the lawn and in a flourish make a row of four cartwheels. I am free. And I can return to the house half an hour later restored as a human being. We cook supper and play with the kids and endure Georgie's shrieking and after bath time, Michael sneaks back into school to tidy his desk and bring home work for tomorrow.

He is an angel. I think we can survive.

Phee came with a small cluster of giggling girls. It was the secondary school's activity week and Michael had offered a week of farm activities. We had racked our brains for likely entertainment. Certainly mucking out the goat and pig and cow sheds; shovelling shit always being a thrill to the uninitiated. Then egg collecting and hay distributing, and poultry feeding, if the schoolkids were able to come early enough or the ducks and geese are able to contain their hunger and cries of starvation late enough. Fruit picking, raspberries and strawberries; beans for picking – weeding, watering, splashing about in the river?

The planning was steaming ahead, brightened by an idea to ring up Louise and ask to borrow her newborn piglets for bottle feeding. The outline for days on our farm was sent into the school and Michael was surprised and horrified to see huge numbers opting for 'Life on Our Farm' – and most of them small girls from the first year.

Eleven-year-old Phee and her two friends turn up in wellies and jeans and have walked down the hill in only a few minutes. They are up for anything, they say. Assorted clusters of others arrive and the farm mayhem begins. Michael and another teacher take groups off to get dirty in the shed, and with baby Georgie on my hip, I hustle a group to feed desperate poultry. Within minutes there are girls running about mad with love for cows, goats, ducks, geese, and pigs.

Gradually the day takes shape and the baby piglets, in a cardboard box of straw, are nursed and cuddled and the goats are groomed within an inch of their lives, and the big pigs gleam after a scrubbing and all the girls are soaked and filthy. The sheds are vaguely mucked out and there are plenty of jobs left for tomorrow.

We are surprised to see Phee and her two friends turn up again the next day and they tell us proudly they have booked in for every day.

After the week of 'Life on Our Farm', Phee comes down with her parents to ask if she can come more often. With a flop of blonde hair, and a smile that splits her face in half, she takes over lugging Georgie around, feeding the hens and putting the goats away; with her friends Helen and Suzanne, they screech with laughter, looning around with our girls, playing in the fields and the hay. They are a gift to pressured parents.

When I am inches from screaming or throttling, I know that either Michael will soon be home or Phee and her friends will stumble in. Together they make crazy suppers for my two, of blue drop scones, or runny cakes and fish fingers; together they pile into the spare room and sleep over, calling it babysitting, but it seems like partying to us. It is simply the best arrangement ever.

Chicken pox

I have a wailing baby because she wails all the time, and now I have a sick toddler, sick with chicken pox and a mighty fever to go with it. It would seem reasonable to dose the sick one with paracetamol, but I can't get her to swallow it.

I have a ferocious technique for dosing goats and cats, and I consider the cat method for Morwenna. Roll the cat up in a sausage of a towel, which keeps angry claws muffled, and tuck the bundle under an arm; use one spare hand in an arm vice to prise open the mouth, and the other to lob in a pill, and then hold the mouth shut and stroke the throat so the pill has to be swallowed.

But she is lying listlessly on the sofa and I try to visualise rolling her up and tucking her under my arm. Plainly she is too big. I could load her down with towels on the sofa and prise open her mouth, but she is looking at me and this all smacks of torture.

I try reason. 'This will make you feel better, just a small swallow of this pink goo, and afterwards you can have an apple cut-up small.' She whimpers and the wailing one is off at full blast, so I pick it up and jiggle it on my hip and try to continue the bribery and persuasion. 'Please be a big brave girl and do this, so I can tell Pa what a good, sensible big sister you are.'

She shuts her eyes and seems to close in on herself. I put down the wailer and feel the poxy one's forehead, which is burning.

OK. A different system is required. I lie her on her sheepskin, peel off her clothes, and I wring out a wet flannel and dab. And again. And again. This is straight from a Victorian novel, and I shrink from the knowledge that Victorian illnesses take weeks, until there is a midnight crisis and then the pale and panting sick one opens her eyes weakly and . . .

By the time Michael has come back to a home of strewn facecloths and damp towels and wailing, the sick one opens her eyes and gives him a brave smile. He scoops her up and they settle on the sofa to read *Each Peach Pear Plum*.

Acorns for pigs

As Morwenna and I push Georgie in the pushchair along the only smooth lane for miles, with Georgie swinging her legs and singing a babbled, wild song, we see acorns scattered wide along the tarmac, squashed by tyres, but many whole, and Morwenna tells me we should collect them for the pigs, Boosey and Hawkes. Where did she get this idea? It's a good one.

We swing right round and romp home for collecting bags. Then what a fine frenzy of gathering, Georgie needing persuading they are not for her to eat, Morwenna insisting on selecting only the perfect specimens and me picking for speed. Georgie sings a celebratory pig song all the way back and we carry our treasures triumphantly into the pig pen. They gobble them down with all the right smacking of lips and snorting in appreciation and we stand cheering them on.

Killing ducks

We have to do this. We are farming, no-nonsense farming, and we eat our own home-grown meat and that includes ducks.

We know which are the young and tender ones, because we have a system of marking the new hatchlings with an ankle ring. Green-ankle creatures, your time is up. There is no way that one can kill and then eat straight away, so it's easier to have a slaughter, a massacre, and then pluck and draw and freeze, so the meat is remote from the killing. So today is duck-killing day and Michael is miserable.

I try to talk him into feeling better about it, but the problem is that Michael and the ducks are friends. I take on the task of keeping the green-ankled young ones in the duck shed and letting out the rest. That takes a bit of time and cunning. Then we are left with a collection of four ducks in execution row. Michael comes in and catches one and takes it out into the yard, where he has laid out an executioner's block of wood and the sharpened axe. It is my job to hold the duck onto the block, thinking of Anne Boleyn, while Michael delivers a mighty blow and severs the head.

He hates it. He says that, 'I hate it, I hate it,' and as I hold the body on the block, the duck turns its deep black eyes to look at him. I am looking away, as I don't want a spurt of blood over my face. I hold the body still as it spasms and the blood gushes onto the concrete. When it is still, I put it to one side and we go into the shed for the next duck. Four down, Michael throws down the axe and staggers off to wash.

He then disappears entirely to restore his soul and I hunker down in the back yard to start plucking. It will take a while, but it's easier when the bodies are warm.

Igloo day

Snow arrived, in fat, floating flakes that settled thickly on the ground. Snow just doesn't happen here – this is Cornwall. It started yesterday when I was in town shopping and I walked on, gazing up at the sky, letting the flakes drift down onto my tongue, and I bumped into a lamppost. I am a fool for snow.

Today that snow quiet had muffled all sounds in the valley and a whiteness was rising upwards into the bedroom when we woke. A day for snow play. Our small girls were surprisingly willing to eat porridge to fill them for a snow morning and were pressed into warm, all-in-one suits and double socks in boots, and we piled out to pat snow, pack snow, and pile snow up into fabulous creations.

Within a few hours, a collection of people had arrived to play; tobogganing on our one sled and some sacks and a few tin trays. Some lads we didn't know took up position further along the field with a terrifying sled assembled out of rusting corrugated iron, which carried four at once in a high-speed descent, tumbling them out halfway with miraculously little bleeding from the savage edges.

Somehow an igloo project got started, and Chris, an ex-pupil of Michael's, was the engineer in charge of cutting blocks and ordering the willing team of builders. I took some exhausted small people in for lunch and Georgie for a sleep and when we returned to the igloo it was well over halfway completed. There was a mounting time pressure to finish before darkness and the builders had carried good blocks of compressed snow from further afield, staggering back before the snow collapsed. They heaved them up and curving inward, always inward.

I missed the last block, and have no idea how it was reached or pushed in, but amazingly there in the field down by the river stands a fine igloo and we squeeze in to sit in a huddle, our faces rosy, ecstatic with our success. I like to think we could have lit candles or drunk tea or settled down there for a night of stories; but at least it will be there in the morning.

Life in an igloo

Morwenna hurls herself downstairs to get to the igloo and I have to restrain her and post bites of breakfast in, squeeze her into snow clothes. She hauls Michael out, but not before commanding me to bring a feast. I watch them ploughing through snow, leaping and jumping, and it looks very fine and Georgie and I hurry our preparations. I pack a bag of flapjacks and apples and grab a battered old cushion and we follow the trail in the snow.

Inside the igloo, family life is a very close affair. Michael's knees and mine are tangling and the girls are squirming about, delighted at their snow home. The sun on the snow walls makes a luminous glow, and basking in Arctic sunlight we have a small picnic. No, we can't sleep in here, not even on cushions.

The igloo lasts, sagging and drooping as the days go by.

Making kindling

There is a scattered collection of small branches and scraps of wood, kindling for the fire, and it is my joy to gather them and bring them in. Many are shards from bigger logs and some are pitiful twigs or rotting splinters from old planking, but they are all ideal for starting a fire; the only factor of judgement is size. Today I have a pile of too-long rotten boughs and I shan't gather them up and put them in the pile for sawing, because I have my method.

I put the short end on a step or granite block and stamp on the wood in the middle and it shudders and collapses into kindling pieces. If the piece is too hard, then the stamping is futile, my foot bounces back and I discard the bough or plank for sawing. I judge by size and weight.

Today my method lets me down. I drag a too-long and too-whippy branch to the stamping place and stomp on it and the straggly end swings back, leaps up and hits me on the cheek with such force that I have to wrench the stump from my cheek. I reel and sit down and staunch the blood that courses down.

After a few minutes, I stagger my way back into the house. Michael is marking schoolbooks and looks up horrified at my gory face. He springs up to attend to me and watches as I splash the blood off my face and then looks at the hole in my cheek.

'How on earth did you manage that?'

'Jumping on kindling,' I reply. He sighs and informs me I am a complete idiot.

Seed catalogue

The enticing pictures and the horrific expense. We are careful with the seed catalogue, trying not to get over-enthusiastic, sharing the expense with our friend Liz with whom we can split many of the packets of seeds. It's a winter activity, poring over the catalogue, dreaming of trugs full of delicious vegetables, oblivious to the weeding, the planting, the hoeing, the thinning, and the fight with slugs, butterflies, snails, and disease.

'Jerusalem Artichoke,' says the glossy picture. 'Easy to grow. Fresh and crunchy with a sweet nutty flavour. A novel winter alternative to potatoes. Foliage grows up to 3 metres high.' Why would one be so seduced by the first sentences that one failed to spot the hidden menace in the last?

Our tubers arrived, sweet and nutty, and were planted out like potatoes and they thrust up leaves and more leaves and then even bushier leaves. We harvested them, sweet and nutty, sweet and farty too, and before we knew it, they had totally claimed one whole section of the garden. Nothing else can grow there, Jerusalem artichokes are in possession, sweet and nutty, and no matter how hard we try to dig them all up, they shoot defiant green flags of unmistakable artichoke leaves.

OK, we say, you can have this part of the garden. You win, we eat.

Sprouts

I have put on my heavy-duty black coat that comes over my wellies and added the yellow sou'wester. I am completely waterproof, which is just as well, as it is raining horribly. The animals don't need to be called in, they are all in bed in protest at the weather and I bring their feed bowls hastily, so the oats don't turn to cold porridge as I shunt from barn to sheds. The ducks aren't in, but it's early, so I turn to the vegetable garden to see what I can forage there for supper.

I have picked all the leeks already and the cabbages were brought in over the winter to hang in the milking shed, dangling from the rafters like decapitated traitors. All I can expect is the divine promise of some early purple-sprouting broccoli or sprouts. The broccoli is a wish unfulfilled. So I turn to the sprouts, which look thin-stalked, mean, and with pathetic little bundles that are the sprouts.

At the first plant, I pull the knobby sprout that looks the biggest and it turns to rotted squish in my fingers. All down the stem the sprouts are either rotten or about to be rotten, with the outer leaves turning grey. I grit my teeth and pull them off, thinking that I will be able to cut off the bad leaves. I'm bending over the plants and the rain is tricking down my neck.

It seems a better idea to find a more promising stem of sprouts, uproot the whole plant and do my picking in the dry of the conservatory. I tug hard and the top leaves shred through my fingers, but a second tug lower down wrenches the stem free, and I shake the loose earth off and rinse the root with the hose. The ducks and geese are up and ready to be put away, so I shut them in and hurry back to the house with my sprout stem.

When I have shed my soaking outer garments and wiped my rain-spattered specs clear, I have a better look at the sprouts. They are going to take some transformation before they are fit to eat. More than half are soggy with rot. Morwenna comes out and wrinkles up her nose. 'What's that smell?'

'Sprouts for supper,' I say, and her look tells me that I might be wasting my time.

Edward admires the countryside

We see Edward in London, child of Michael's oldest friend; the girls love him.

Edward is allowed to visit; his mother finally permits it. His father arrives with an inventory of the five-year-old's clothing. Edward makes a slow, careful passage from the car. He walks like a toddler still, as if his head is still too heavy for his body and needs thoughtful management. He goes down the granite steps and he stands and surveys the garden.

Our children bounce out of the house and rush around him like puppies and he receives their affection in a kingly way and continues his appraisal. He toddles down to the edge of the garden and gazes over the meadow below and makes his pronouncement with a wave of his hand, 'Oh, what beautiful countryside.'

Morwenna looks baffled, but can see that it's a positive remark. The adults collapse into stifled giggles and Edward resumes his royal progress over his new land and walks into the house. Morwenna wants to drag him under the stairs to the playroom, but he stands firm to assess the kitchen. With his feet planted squarely in the middle of the long kitchen, he stares at the bright yellow sofa and the curtains of the same colour and he gives a little sigh, 'Oh, what a wonderful sofa.'

Edward, you tiny old man, I love you. You have said the right things. You will always have a place here.

Chapter 4

Flood in the bathroom

Michele brings her children to play. She and I are sitting at the kitchen table drinking tea and Georgie and Lorna are playing under the stairs and baby Perle is obligingly asleep, so all is well with the world. There is, of course, much chatting between us and I daresay time slips by and we stop noticing the two girls' small noises of playtime, which have, in fact, stopped.

This is because they have gone upstairs on a mission to the bathroom.

It is only when a faint, regular sound breaks through our conversation and we identify it as a drip splashing fatly on the kitchen floor, look up and see a steady welling of drips forming on the ceiling, that we hurry upstairs. From the top stair, I can look into the bathroom and think that I see Georgie floating by on her potty. The entire bathroom is awash and water is cascading out over the top of the bidet. Lorna and Georgie are seemingly unaware of this watery problem.

I turn off the tap and pull out the stopper and shriek to Michele to run downstairs for a dustpan to scoop up water. We act as a fast and frenzied team. We lift water and hurl it out of the window, scoop hurl, scoop hurl, and then begin a blotting process. There is a stack of bath towels in the cupboard and we throw each one down to soak up a full wetness and then chuck it out of the window. By the eighth towel, the flood is being pushed back.

The girls have enjoyed this enormously, paddling about in soggy socks. The bathroom is having a spectacular spring-clean, and little bits of fluff have swum wetly from hidden corners under the bath and behind the loo.

We leave the windows wide open and go out to the yard to gather in the sodden bath towels from below the window. We have lunch. What I do know is that I shall not be telling Michael about this. What I do not know at this point is that when Michael and I sit to watch the *Nine O'Clock News* later that day, a whole strip of wallpaper will slowly peel back and slide down the wall as I watch in horror. Michael inexplicably does not see it.

After the news, I lure him away and rush back to persuade the wallpaper to hang up again, shoving it up with a broom.

Glass slip

What should be in a conservatory? A sofa. A dining table and chairs? It's easy to find those. We have an old rattan sofa from our Malaysian honeymoon and we have a white-painted occasional table with four little white chairs from a junk shop. To each side of the granite floor slabs we have left a flowerbed, rich in compost from the cowshed, and these we have filled with an abutilon and geraniums. We assembled a swing from the rafters for the baby. And a painting. And a box of sand.

So the conservatory began and now it has flourished into a wild jungle of plants and a collection of wellies and the sandbox for the girls. There is a cluster of sand toys and assorted abandoned dolls, pushchairs, building blocks all around.

Today one of the giant triangular slices of glass that form the five-sided front of the conservatory slid out of its holding, pivoted on itself and crashed headlong, point down onto the granite floor by the sandbox. The crash was ear-splitting. The splinters and shards of glass were everywhere. To be in the conservatory at all would have meant serious injury, to be under the glass would have been death. Our little girls were there only an hour before. I am trembling.

Rosetta A38

First thing this morning, Louise rings to say can we take our cow back, the one she borrowed? Rosetta, our only cow, idle, middle-aged and greedy, was borrowed a month ago to fulfil Louise and Chris's farm-holiday brochure promise that their guests could help with the milking.

Rosetta's departure had been swift and straightforward. Chris had rumbled into the yard with a battered horsebox and Rosetta had been persuaded into it. The horsebox had made a worrying departure, squealing and grunting along our muddy lane, and by the time it reached their farm five miles away, the rusting chassis had disintegrated, and they all came to a jolting and permanent stop in the driveway.

Rosetta had been a huge success with the visitors and now Louise was drowning in an excess of Jersey milk and they had run out of oats and hay. Could we pop round and collect our cow?

This requires a full family conference. We sit round the kitchen table with big mugs of tea and wonder how to do it. We have a small van, too small for a cow. Colin, Jack's son, so wonderful and neighbourly, is busy with planting. And we don't have a towing bar. Liz drops round and considers the problem. Morwenna offers her small child's logic: since Rosetta could walk up the hill to Colin's bull, why can't she walk home? Well, yes, why not?

We get maps out and trace a back route along a hardly used lane, through the wool grading yard, over the bridge, through Boduel and down. It could work, but there is one problem: we have to cross the A38, at a point on the busy dual carriageway where cars start bunching up to crawl up the hill into Dobwalls. How to cross with a cow? How fast can we persuade Rosetta to go?

Further mugs of tea are needed. But we have a plan.

Phone calls are made, promises are given, details are thrashed out and we are to meet at 9.30 a.m. tomorrow. A Saturday. I have worrying dreams.

One by one the team assemble. Michael as farmer to lead a cow. Me as support, troubleshooter and pusher of pushchair, Georgie in the pushchair plus drinks and her sheepskin, Morwenna in her rainbow-stripe jumper; we are all very brightly dressed for high visibility. Rae arrives with baby Peter in a buggy decorated with ribbons. Liz takes command of our van and in it lies the signal flag: three bamboo canes lashed together with a large red rag attached. She drives off to wait at home for a phone call from Louise to tell her to set off and wait at Dobwalls in a lay-by. The timing of this is going to be key.

We drive to Chris and Louise's farm with the car holding important bits of equipment: a halter, rope, a bowl of oats, a red skirt doubling up as a second red flag, two baby buggies, and Morwenna's recorder.

At Trengale, there is readiness. Rosetta is very keen to have the halter in exchange for the oats and starts off the journey at a brisk pace, so we wave a hasty goodbye to Louise and follow Michael and Rosetta who are already well on their way. We are a jolly group; there is singing and recorder blasts and chat and the morning is sunny. After about a mile of dappled sunshine and no interruptions of any kind, Rosetta gets bored and stops to eat wayside grass and we are all glad of a break.

We sprawl on the bank amongst primroses and the small people skip about. As we laze about in idle chatter, I wonder if Liz has set off in the van. We can't afford to linger too long and so I push the team on. My heartbeat is beginning to race at the thought of all that can go wrong, like a cow and family stuck at the roadside watching cars roar past, or children or cow scattering in panic and getting snarled in traffic, or other worse horrors. I can hear the A38; we are getting closer.

At the approach to the big road, we have to climb steps to a gap in the stone hedge where we will cross. Michael goes first, cautiously, but Rosetta is keen to follow. We all cluster on the pavement to consider the task. The choice of Saturday has been intentional. Saturday holiday traffic usually means a stop-start crawl up to Dobwalls, the easier for us to dodge through. But today, of course, the traffic is flowing smoothly and rather fast. No one seems to have noticed our strange group at the roadside.

We strain to get sight of Liz. The plan is that Liz will be driving slowly, slowly down from Dobwalls with the bamboo and red rag flying, with frustrated drivers behind her forced to a snail's pace. She can then stop and this will enable safe passage at least to the stripe in the middle of the road, at which point almost-crawling drivers heading to Dobwalls will see the situation, cow, family, etc., and let us pass.

No Liz. Rosetta finds some grass and we wait nervously. Suddenly Liz appears at the top of the hill, her approach very slow indeed. Cars are hooting, but at least the double white line prevents maddened overtaking. The road is clear from the right and I take my flag and stand in the middle of the road and wave it to show that something is amiss. Hopeless. Car after car drives past ignoring me. How? Why? A crazed woman waving red stuff doesn't merit a second glance? I move further into the fast lane and a few cars swerve to avoid me and finally I get them to stop. The River Jordan has opened.

Michael has his chance and leads Rosetta over, followed by buggy, pushchair and Morwenna and Rae. We are over!

From there it is an unhurried pace. I am shaky with adrenalin, but everyone else is completely relaxed. There is a stop for snacking and recorder playing, a chance for Rosetta to munch on hedgerow treats, and some climbing in and out of pushchairs. We dawdle back through Boduel, the wool yard, the back lane. Rosetta, welcome home.

Wellies in the washhouse

We have a sprawling number of houseguests, lolling on sofas and reading newspapers and I am thinking that they might prefer to be helping me on an ambitious garden or farmyard project. It's not that any project is particularly pressing, it's more that I am not reading the newspapers and there is no room left on any sofa.

I scrunch about thinking of best projects and announce a few to gauge a response. There is only a vague grunting, so I declare that we are going to clear winter flood debris from the riverbank as a rescue mission to stop the alders and other trees on the bank being undermined by currents and blockages. My tone of voice and the Samaritan aspect of the job seem to have the right effect and people shift about and start matching their clothing to the task.

In the boiler house, I have located some dusty and spidery wellington boots and bring them out triumphantly. There are three pairs of size seven, two of tens and one of eleven and a vast array of small children's boots. Three potential helpers look affronted and say they won't be able to come down to the river wearing those. The rest, sensing an advantage, start to dither about, so I have to act fast.

'It's okay,' I say, 'I'm going to drive up to Liskeard to get the world's best pasties for lunch and next door is the world's best wellies shop, so tell me your size and I will buy you a pair.' I think I have been spectacularly clever, as there is almost a suggestion of no lunch for no wellies and quickly enough I have a list of required sizes. The lunch and the riverbank project are both a success.

Michael and I put up a special shelf in the washhouse for those wellies with the size written in white on the outside in correcting fluid. We have something for everyone, for evermore.

Broken-down tractor

Broken-down Old Tractor, Riding on You.

This could be a song, lyrics by two small girls and their friends, and the words and the melody keep changing.

Halfway down the drive. Enough of a walk for wellies and a coat. An old Fordson tractor lies buried up to the wheel axle in overgrown weeds. It has no cab, but it has a fine moulded seat, strong and squeaky pedals and a great big steering wheel, which jiggles from side to side. We hack back the weeds at regular intervals. Granny can get up there, too. A whole motley crew of chldren can clamber about without too many squabbles and falls. It is a space rocket. It is a pumpkin carriage on its way to the ball. It is a ship. It is even a tractor.

We are sad today. It has gone and all that remains are deep gouges in the verge. There is talk about where or how it can have gone, but I know with a weary heart that it has gone to the big old tractor dump, probably to Squeaky Orchard's in Dobwalls, to offer itself up for organ transplants. We should have bought it for ourselves, to lie in the verge forever.

Kingfishers

I know there must be kingfishers along our river. It is the most perfect kingfisher place, with low branches across the water and a good supply of small fish. The water runs fast and waves the weed about like mermaid hair, and the glint of stones makes the bottom seem shallow. The roots of the alders are long fingers into the water and they grasp and snag twigs and flotsam, but it is a fine, sinewy river with hospitable quiet bays where the ducks paddle, and it has fast flowing patches too, where a child's wellie boot can be filled in a second.

So why, when I watch and watch, do I not see a kingfisher? This river is loved by herons and our ducks and geese, but there is room for more creatures. I can't believe that our noisy flocks would deter a shy kingfisher. My eyes scan the river every day and I do not see a kingfisher. I know there must be kingfishers along our river.

Georgie's beaker

Georgie. In her wellies and padded splash suit, holding my hand and waddling out to the yard to watch Pa milking Rosetta. There is a special mission here. She has decided that her breakfast will be warm Jersey milk fresh from the udder.

Michael looks up somewhat surprised to see us, since we were sitting ready to eat breakfast a few minutes ago. 'Pa,' she announces, 'I need some milk in my beaker.' He looks at the two-handled, orange plastic beaker doubtfully and explains that he might have difficulty aiming it in, but she just gives him that look.

His grip on one of Rosetta's teats is firm and he squeezes carefully and a strong jet shoots down into the beaker and splashes right up. The next squeeze manages to stay in the cup as he holds it lower, and with each pull on the teat, the milk foams and rises in the beaker and he then pauses and hands it over to Georgie.

Solemnly she takes the beaker, eyes it as if it's a pint of Guinness waiting to settle, and then pouts a little as she sips it. No words, just an affirmative nod and a pleased look in the eyes. She drinks it down straight. With a little skip, we are back towards the house to sort the rest of breakfast and going to nursery.

Cement

I have been wandering about with an enormous hired video recorder lodged on my shoulder; I am David Attenborough, recording farm species, recording weekend guests at play, recording life, when I stumble upon a slight problem.

Morwenna, Zafar and Edward, my trio of free-range children, are in the hay barn and they have found our sack of cement powder and are busy mixing it with twigs and a dribble of water from a broken flowerpot. I have snuck up behind them, and I continue filming and remonstrating at the same time.

'What are you doing?'

They leap up and assume instant expressions of guilt or blank denial.

'Nothing,' says one. 'It wasn't me,' says another, and then my truthful child says, 'We are making cement.'

'Well, you had better do it properly then,' I reply. 'Morwenna, get a bucket of water from the milking shed. Zafar, get a shovel. There is sand in the sack over there.' Ed is left alone with me, so he looks firmly at the ground and shrinks a little into his duffle coat.

Clarissa, my oldest and best friend, appears and calls her son Zafar a clot, but the mixing is going well. They have shovelled out some paltry amounts of dry powder and are splashing water into it.

I can't resist the impulse to give instruction, so tell them to get out more powder and equal quantities of sand, less water and to slice and turn the mix. They get the hang of it quickly, taking turns and then we get a wheelbarrow and the morning becomes a cementing sort of morning as they trundle their mix into the yard and splosh it into holes and flatten and smooth it out.

Good work for six-year-olds. I pick up the video camera and resume observing life on the farm.

Morwenna's class diary

Michael and I are perched on impossibly small school chairs attending a parents' evening. We listen contentedly to the teacher's description of the current project work, times tables, reading plans, etc., and then are given a folder each to admire our children's work. We flick through the artwork and then read with interest some of the statements in Morwenna's News Book.

Here we learn all sorts of surprising things about ourselves and our lives. My daddy cooks. Really? We picked apples. My mum mends the car. Did I, when? We killed nine chickens and we put some in a cage. There is an alarming drawing to go with each of these statements, lumpy hens dangling on a pole, vast buildings, someone's legs sticking out from under wheels, and a curly-haired girl with a basket under a frightening apple tree. We wonder what her teacher thinks.

Gazing neighbour

Robert is standing by his five-bar gate looking over his land, into the field that is the twin to ours, the other side of the bridge, facing north towards the waterworks and the old manor. His arms are resting on the top bar and his head is tilted to one side; his weight is on one leg so that he is faintly lopsided. He is wearing a flat tweed hat and a rough farming jacket and his wellingtons are high on his short legs.

We are on our way to the bridge to throw sticks into the swirling current, but we stop for a neighbourly chat. The little girls scatter and look for wild strawberries in the hedgerow and I join Robert in his gate-leaning. I expect him to be narrow-eyed, admiring his cattle, but the field is empty. 'Just look,' he says, 'isn't it beautiful? I love this river and the meadow.'

And we pause there, together, for a moment.

Butter making

The girls are standing on chairs, so they are high enough at the table to make butter. They are swathed in a tea towel each as an apron and I have scrubbed their fat little hands. Today's two litres of cream have already been whipped in the food mixer, turning from thick to frothy and then finally succumbed and separated into fat particles and buttermilk. I pour off the buttermilk to make into drop scones later, and rinse the butter granules in a sieve under the cold tap.

Each of the girls has a bowl of butter granules, and a pair of old-fashioned wooden butter hands, which look like grooved, flat paddles. There is a bowl of water on the table, for rinsing the butter hands. The dairy job is to gather up some butter and press it together and squeeze, working the butter into a shape, a pat. I sprinkle a small amount of salt. There is chatter as the girls squeeze out the remaining milky water, and they roll the butter on the paddles. When they are satisfied with their butter shapes, a ball, a blob, a squarish item and so forth, they turn them out onto sheets of greaseproof paper and wrap them.

I watch them mothering their butters and wonder how a boy would have tackled this little project. Just fine, I reckon.

Comet time

Each night it is there, high up but dazzling. A real comet, for us in our lifetime. We are in awe of it, visible as we walk briskly in the winter night from swimming lessons, or out to check the animals, or leaning far out of a bedroom window.

We are accustomed to pointing out Orion, Ursa Major, the Pleiades, Cassiopeia, and they are our friends of the winter night sky, but Comet Hale–Bopp is a rare and wondrous thing, with a fierce brightness and a fine tail trailing into the black. We look for it every cloud-free night, longing for it still to be there, afraid that it will have soon spent its journey with us.

'Look, look, there it is,' we tell ourselves, richer by a comet.

Freezing

This whole house is freezing. Michael says we should wear extra jumpers. I am wearing a long vest, two long-sleeved T-shirts and two jumpers. My feet are cold and I probably have chilblains and the radiators won't come on for hours.

Go and chop logs, light a fire. Get the kids into warm splash suits and mittens and bobble hats and we can all chop logs and carry hay and run down to the bridge and throw sticks in the fast flow and then huff and puff up the hill and go off sideways into the woods and jump in puddles and come back warm and with rosy cheeks and peel off the big clothes and make pancakes and light the fire in the sitting room and sing songs.

It's a plan.

Colin babysitting

There is a loud thrumming of tractor tyres in the drive and the slippery body I am drying has wriggled free to rush down the stairs to see who is there; I catch it and press it firmly into pyjamas. We can hear Colin before he even gets into the house, greeting Michael along the drive and we are in the kitchen before them, ready to say hello.

Colin is going to babysit and this is alright with both little girls, who admire him greatly. His booming voice and clinging smell of the cowshed, his nails bitten to the quick and engrimed with farmyard do not deter their love – he is the driver of the vast tractor, the dependable body when it comes to herding escaped cows and he has the sweetest smile.

He has driven down in the tractor despite being only fourteen years old and has brought with him his evening's entertainment, which he is busy spreading across the table. It is his arsenal of weapons, a rifle and a shotgun and an assortment of cleaning cloths and light oil; the plan is to strip down his guns and oil them and fix them up again.

Morwenna nonchalantly places her pink doll on the table next to the guns and demands a bedtime drink and a story before we go out. I persuade her against getting Colin to do the story on the yellow sofa and wheedle her upstairs with Georgie for stories and bedtime. 'Say good-night to Colin and be good.'

He roars out a cheerful parting, 'G'night, little maid,' and when we slip out of the house he scarcely looks up, so intent on his weaponry.

Guinea fowl

They are all on the roof making one hell of a din. Have they seen leopards? The guinea-fowl experiment is both charming and tiresome. Their various disadvantages are their smaller-than-a-mustard-seed brains, their peacock screeching and their refusal to be domesticated.

We were not aware of any of their bad habits when we bought a clutch of eggs, on a whim, to incubate.

The tiny hatchlings were exquisite, tiger-striped and eager for life. They were reared under a motherly red lamp in a wine box in the tool shed and fed with chick starter crumbs. The cat tribe was at war with them from the start and we knew it. There was an excessive display of hunting body language, all those haunches low and slinking along the ground, and a maverick cunning in trying to sneak into the tool shed.

We put the week-old hatchlings outside in the rabbit hutch, whose chicken-wire walls, sheltered housing and open grassland seemed ideal, and then watched them make an escape within seconds straight out through the little holes in the wire, squeezing tiny bodies into impossibly small gaps. The cats became very excited and it needed some energetic catching of fluttery guinea babies in order to preserve life. We posted seventeen back in the wine box and had no idea where two were hiding.

Archie, beloved mackerel-striped cat of our hearts, knew exactly where one was and leapt into a bush and came out with it crumpled in his mouth. I shrieked at him as he ran off with his catch. It wasn't too hard to find the last little chick in some long grass and he re-joined his friends in the tool shed.

At Cornwall Farmers the next day, I found some fine-woven, black nylon netting, the mesh too tiny for any escape and I stitched and stapled a small run to the wall and there the guinea babies grew up into ugly teenagers, much used to menace from Archie and Bartholomew, who would line up and stare.

They quickly learned to fly and screech; within weeks they were too frantic for any further lingering in the netting run, and given freedom they chose life in the orchard, perched in the branches of the apple trees yelling at everyone and anything. They then flew up into the very tall pine, and from there the top of our roof was easy.

At five in the morning, the distant call of the cockerel from the yard is a pleasant bucolic call, while from inches above our heads, the reply of the guinea fowl sounds like hell in torment.

Jane Ledstone

We build the greenhouse; windows with controlled openings, a sliding door, a brick foundation and a million wheelbarrow-loads of well-rotted compost.

We have some plants lined up ready to go in before we spot the problem. There is no central solid walkway and the compost is still loose – the first step Michael takes in the greenhouse has him sinking halfway up his wellingtons. We need to lay a path.

'That's fine!' I shout and dart off to the big barn, where I am sure I remember seeing paving slabs and I come back with two. They don't go very far.

'We could use the gravestone,' I say. 'It seems a good use for it.'

We had stolen it from the house we rented, outraged to discover that our clergy landlord had laid a splendid patio of slate rectangles that turned out to be gravestones.

Together we manhandle the heavy slate into the greenhouse and lay it down properly, face up: 'To the Memory of Jane Ledstone, Dearly Beloved'. She lies remembered in our greenhouse.

Wedding fever

We have wedding fever. Ever since bridesmaid duties were bestowed, we have been in a swirl of extravagant fantasies. I made the cream silk sailor frocks, bought the white ballet shoes, but the bride produced the crowning glory: the flowered coronets and posies. They are used every day.

The dressing-up box has yielded veils and frilled and flowered dresses and so Morwenna and Edward have plighted their troth in a top hat and flowered coronet. Georgie is keen to have a husband and today has been squirrelling about in the dressing-up box humming busily. She is already in full bridal garb. Her veil keeps falling and obscuring whatever is going on in the toy zone under the stairs.

She emerges triumphant and comes into the kitchen dragging the inflatable Pink Panther by the hand. He is dressed in a jaunty blue sweater and some silk harem pants. She arranges him by the kitchen table and asks me to marry them. I put on a solemn voice and tra la la the wedding march and then ask each if they promise to have and to hold in sickness, etc. Georgie gets very bashful and declares she will.

I resume my pasta making and Michael is marking schoolbooks when a loud shout is issued from the loo. 'Pa!' . . . He goes to attend to her and returns to say she is holding hands with the Pink Panther on the loo. When she comes back, he has undergone a costume change, now sporting a tie and a T-shirt. 'Why the tie?' I ask.

'Oh,' she shrugs carelessly, 'now we are on honeymoon.'

Chapter 5

Massacre

A big fuss in the yard last night.

Michael comes stumbling into the normal, cosy domestic scene: cornflakes and Barbie dolls, and gasps, 'There's mess and feathers everywhere, a fox has killed all the chickens.' We are all out in the yard in a moment.

There is a frenzy of feathers and slumped brown bodies with gashes and gaps of darkening blood. Two hens are running in a crazed way, but the fox has fled the scene. Morwenna counts and recounts the dead, the dying, while Georgie clings to me, trying to understand what she sees, 'What is it? What?'

We put the dead in a barrow. The two remaining living hens flap into the field.

Once we had a lot of hens, now we have a lot of bodies. The farmer in me is not going to put these to waste, so I assemble a chopping board and knives and set up a field butcher's shop. It is quickly apparent that only a slip of breast meat is worth salvaging. I slit along the breastbone, easing off the skin and filleting away the morsels of meat, still warm. Morwenna leans into the barrow and prods various body parts, asking, 'What is this . . . that, this?'

I willingly give up the meat processing to open up a hen and show her the intestines. We find an egg ready to be laid, and a series of almost-eggs in a chain along a tube. We look at the heart, a small purple knot; the crop from the neck, which is a membrane bag stuffed full of grain and grass and unidentifiable edibles; the gall bladder, the lungs; each and every fascinating thing.

New hens

As the fox has taken all but two hens, it is time to replenish the stocks. I can buy point-of-lay young ladies, awaiting the birth of their first egg, but last time I did that I had to wait a very long time for any eggs. The hen farm does not have any yet, but they could offer me retired hens, who might lay.

I took Georgie and Morwenna and a cardboard box to Pensilva, to the hen farm to see what retired hens looked like. A matter-of-fact kind of guy led us to a waiting place and took away our box, returning with six ordinary, brown hen-like hens. This wasn't turning into a very exciting project, no thrills or bartering, just hens, but we paid £1.50 each hen and went home for tea.

Apparently good housekeeping practice for hens or indeed any new creature is to rehouse them in the dark. So we waited and then took them to the designated hen zone and one by one reached in their carrying box and picked up a snoozing hen and put her on the wooden ledge.

In the morning, we were keen to greet the newcomers and both children came out with gifts in the form of left-over toast and a bowl of corn. The usual frenzy of poultry noises was in full spate outside in the yard when we opened the hen door. Exactly where we had left them were six befuddled hens. They hadn't moved a muscle or a feather. The girls sprinkled corn and toast about in the straw on the floor, but the hens just blinked.

We busied about bringing a metal drinker and agreed that they should remain shut in the barn for the day, so they would know it to be their new home. Still none of them hopped down to eat or drink and we looked at them feeling rather puzzled. Surely, they were hungry? Hunger seemed to rule every other creature on our farm.

Morwenna had the idea. Maybe these hens didn't know that they could come down and eat. Maybe they were too used to food in a trough. Perhaps they were afraid. Or they didn't know how to move off the shelf.

They showed no reluctance to being picked up, not a quiver of protest, not a squawk. Each one wobbled a little as she was put down and then cautiously moved forward towards the drinker, where one by one they got the idea and took dainty little sips. One brave lady even put her head down and pecked at a grain of corn. We stood by watching, fascinated. They were so timid and shy.

At bedtime, in the dusk when we had put all the other creatures away, we found the hens huddled in a feathery group in a corner. 'Hens, you are meant to perch,' I told them and picked them up and put them on the shelf. Of course, it dawned on me, they had never perched in their life. They didn't know that hens did something called roosting, where small skinny claws curled round a branch and kept them there all night.

That night their training programme of how to live a proper life began, with all the treats of freedom and its dangers.

Hens, you scratch for bugs or grains or scraps; you peck at grass and it makes the egg yolks orange; you wander in the fields and make comfortable noises to your friends; you make a dust bath when the weather is dry and you fluff out your feathers to clean them; you will feel like crying out in utter elation when you have laid an egg; you get to choose a nest for eggs and lay one after another until you have a clutch; you might even hatch out some chicks; you watch out for foxes and die of shock and panic when they grab you by the throat and tear you apart.

This is proper hen life. And there is more. I will bring a cockerel into your life, and you will love him blindly. He will treat you well at times, calling you when he finds worms or a good supply of food, but he will have his wicked way with you and then show disgracefully bad manners by standing on top of your submissive crouching body and crowing. It's hen life. Get used to it, grateful hens, better than the life in a cage.

Grateful hens

The hens listened. They are happy old hens, who move in a companionable group, murmuring and muttering to each other. They are pecking in the grass and are probably frightened by the huge world. So far they have managed to leave the barn, the yard and have entered a few metres of meadow. They cower as the ducks race past. Three eggs have been laid on the floor of their barn and they haven't yet managed to roost for the night. Each night I pick them up off the floor and put them on the shelf, but this morning two of them had managed to jump down.

I feel proud of them trying this new life, and love seeing each brave new step they take. The cockerel is coming next week.

Colin and the pram

The girls scatter and play where they will, and I have a lurking worry about the terrifying teenage speed at which Colin drives his tractor down the hill. I warn them not to go out onto the shared drive, but I don't feel sure this is enough. I have asked him, 'Colin, you are way too fast down the hill, what if the girls were there?' and he replies, 'Oh, don't worry, I'd never hurt the little maids.'

But could he stop? Could he? This morning I heard him doing his mad rush down the hill and then back up again in less than an hour. That's it. From out of the toy cupboard under the stairs I drag the life-sized doll that Morwenna nursed for months and stuff it into the dolls' buggy and I carry it out to the edge of drive. I park it with one wheel on the concrete drive and the rest in the foliage by our farm sign.

By mid-afternoon, I have forgotten about the doll in the buggy. A violent squeal of brakes reminds me.

A few minutes pass in silence and then Colin's tractor engine revs up and it makes a slow rumble up the hill. That should do the trick.

Cheese making

It is obvious that in my dairy world of full-cream milk, thick double cream, yogurt, buttermilk and butter, I should also be making cheese. I have a scattering of equipment from a short cheesemaking course I attended: a mould, mats and rennet, and I think I am able to get the kitchen scrupulously clean. So here we go.

Cheese one: coulommier, rather tasteless and dull.

Cheese two: goat's cheese drained through a muslin. Good, quite goaty though.

Cheese three: cheddar, left to age for months and wrapped in a muslin bandage. Delicious and where it cracked it grew blue mould, so we got a hybrid stilton/cheddar.

Cheese four: camembert, extraordinary, stinking to high heaven but tasting delicious.

Cheese five: of our own devising, mostly resembling a pepper-studded doughnut.

Cheese six: unrecognisably anything, stinking and tasting foul.

Michael is the greatest fan, praising them all except number six, which gets hastily offered to the hens and ducks. The girls shriek and pretend to faint and hold their noses and steadfastly refuse to eat any.

'Why can't you make Dairylea triangles?'

Smell of spring?

There is a hint of a gleam on the bare twigs down by the river, a sheeny green barely visible and to me, in a heartbeat, this shouts spring – an end to relentless grey days, a swirl of brighter blood in my veins and in everything around me. Grass, please start growing. Primroses, unwrap yourselves. Dawn, please come before the alarm clock. Hens, start laying. The sap is rising! There are hot cross buns in the supermarket. We must pick daffodil buds and bring spring inside.

Stoat to the nature table

Morwenna is as keen as the next kid to take things into her reception class and there is plenty to choose from. She takes in bunches of grapes, frogspawn, a leaf that has rotted leaving a filigree skeleton, flowers, caterpillars in a jam jar, a slowworm killed by the cat, shells from the beach, some beeswax, and today there is a dead stoat on the mat, which she eyes with interest. It has rich brown fur on its back and a creamy white belly and throat, a small tail, and an exquisite face with a mouth full of razor-sharp teeth.

Morwenna demands a box and picks the stoat up carefully and lays it out on a stretch of loo paper and proudly takes it on her lap as we drive to school.

When I collect her, I am caught by the teacher, who starts off in a positive manner: lovely to have all these wonderful things on the nature table, lovely to have such interest in nature, lovely to live on a farm and experience all these things.

But she is not quite sure that dead animals have a place on the nature table, and considering the other children are less accustomed to farm life, could we restrict Morwenna's offerings in future to plant life?

Judy in the teeth of revenge

Judy has escaped from London, from her misery and disappointment at Richard and his bad behaviour, from term-time and from life, and she will be with us for two consoling weeks.

The children are thrilled and have been clambering over her, requiring books to be read, for her to join them in a tea party, to curl up with her on the yellow sofa for girl time. Michael and I have left her and have taken up the usual collection of Cornish shovels and the digging spike and bits of sacking for kneeling on, and have started a session of weeding.

Judy appears, dressed as if for a smart day out, which is her own inimitable style, and says the girls are busy with the dolls' house and she has come to do weeding. I offer a rougher outfit and boots, but she smiles and says this is her gardening ensemble. She joins in on the onion bed and I move away to thin the carrots.

I look back, because I can hear a laboured breathing. Judy is transformed. She is wrenching weeds out with fury and passion. She is breathless, vigorous, hair wild, hurling clumps of fat hen, groundsel and couch grass into a bucket, hacking at roots and cursing under her breath. Is this Cornish therapy?

New compost boxes

Michael has been completely occupied making a new compost box, following a design in the self-sufficiency handbook. It is huge and has two compartments, one for new compost and one for rotting compost. There are slots on one side for wooden panels to rise as the boxes are filled and the other three sides are made of sturdy timber. The bottom has branches to allow circulation of air, and the top is a lid to which Michael is putting the finishing touches. It will cover both sides and has a mysterious hole in the top.

Judy and I have had occasion to look at the work in progress and make enthusiastic, supportive noises. At the end of the day, the project is declared complete and we are invited to find any greenery that needs to be composted to start the process that will turn waste into black gold.

Michael then surprises us by explaining the hole. In his best scientific voice, he lectures us on the value of nitrogen on the composting process and states the best source of this is urine and then he says that to start the compost off with maximum benefit, we need to give it this blessing. He brings out a stepladder and asks us both to please climb up and wee over the hole. Such is Michael's authority over this project, I start to obey unblinkingly and climb up the ladder. Then I see his gleaming expression and nip right back down.

"No way am I going to squat over the top of that box. Get a bucket."

The Kenwood Chef

My major bit of equipment is my kitchen Kenwood Chef. It is busy every day with cake making, bread kneading, butter churning, liquidising soup, whipping cream, sausage making, and more. Today we may have broken it, showing off.

Visiting us are people that my friend Clarissa and I want to impress, people who annoy with their boasting of geese-rearing and farming wizardry and we have decided to produce a lunch of homemade sausages. I have breadcrumbs, fatty pork, herbs, salt, pepper and a long string of smelly sausage casings, which I keep in a jar in the pantry packed down with salt to preserve them. They are real pigs' intestines. That should impress anyone. They get lots of rinsings before being threaded onto a hollow tube on the mincing attachment.

We mix our ingredients and assemble them on the flat dish above the mincer and push through the mix, watching with great satisfaction as the sausage comes out and we twist the casings at regular intervals. We are so intent on our smug work, we do not notice Georgie sidling into the production line.

The sausage-making system suddenly groans to a halt. I imagine a tough or grisly bit of pork and so press harder on the pusher to force whatever is blocking the mincing attachment. I am pushing with all my strength, when the machine gives a shudder and out of the nozzle and into the casing comes a strange, solid shape. We peer in. I break the casing to get a better look. Here we have a plastic zebra in sausage form.

Thank you, Georgie.

Worming day for everyone

We have hens looking wretched, hunched down, dejected, feathers messy, pools of hideous droppings. Sickness. At the vet's, I am advised that it is an affliction of parasites or worms and everything is probably infected and the hens would show it first. I should treat everything. All ducks, geese, hens. 'It's simple,' he says, 'just put this mixture in the drinking water.'

All the creatures co-exist in a hugger-mugger fraternity and so I think if we are treating poultry, we had better treat four-legged creatures too, and maybe two-legged. I buy the poultry medicine and get the worming 'drench' solution for giving orally to the goats and cows. Buy cat wormer, too. Do I stop by the chemist and get a dose for children? Definitely. Adults? I decide to discuss this with Michael first, but surely he should agree in fairness to treatment for all, himself included.

The programme will involve all ducks, hens and geese being kept in their assorted night-time apartments, getting normal food and treated water in the morning, and being imprisoned inside until the water has been drunk. I have no idea how long this will take.

The goats are straightforward, but more difficult. Their oral dose is in measured syringes and it takes two of us to administer.

One person, me, holds the goat by the collar and backs her into a corner and then leans against her to immobilise her. The other person has the syringe and forces the goat's mouth open by sliding a finger along the back of the gummy mouth and slipping the syringe in and squirting, while at the same time holding the head up, so the goo doesn't slip out sideways. This time we have only seven goats to treat and no bullocks, so it should be manageable.

Yamaha sees me coming in with the syringes and looks at Michael and we can almost hear her say, no I don't think so, not for me and she deftly avoids capture, planting hooves firmly in the deep straw and bucking her head away. 'Try someone else,' orders Michael and I grab Plum and twist her into the corner before she has had time to think about it. Syringe in quickly and she coughs and splutters.

Now there is an air of panic amongst the herd. Michael and I lurch from one side of the shed to the other, making wild grabs, and I clutch Spotty's collar. Manoeuvre. Syringe. We decide to let everyone cool off and calm down, including ourselves, and we go to inspect the duck shed.

The medicine water is in a proper drinker with a narrow rim, so the ducks wanting their morning bath can't climb in. They seem to have finished the water and there is no way of telling if it went in their beaks or they sluiced it out on the floor, but we let them out and they rush mad with joy down to the river. Back to the goat shed, sneakily, calm, several more to go.

Making cider

We have a million apples crashing to the ground and the glut bothers me. When someone says they have a friend who is making cider, I beg to be looped into the plan.

There was a mysterious phone call last night and the apple gatherers will be here today, we are told. All we have to do is gather the apples in sacks and they will be taken away. The reward will be a share in the yield of cider.

Early this morning, thankfully a Saturday, we are all out in the orchard and garden with sacks. We have enough paper feed-sacks for a great many apples, as long as the apples are not too squishy to sog through them. All is optimism and energy for a few minutes until reality cuts in. The apples and the grass they are lying in are very wet indeed. Some faint-hearted girls turn squeamish noses up at apples in advanced state of sogginess. 'Well, leave those ones,' we mutter, but that isn't enough and a search for gloves interrupts proceedings.

Then the wasps are up and busy as early as we are, hiding themselves in half-eaten apples, and almost as soon as we have started the sack work, Georgie gets stung. Is it vinegar or lemon juice or bicarb we must apply? A special nest is set up on the garden bench so that the injured party can observe the rest of us at work, in gloves.

How many wet apples can go in a sack before the sack gives out? How long before the collectors come? How yucky an apple is a good cider prospect? We decide to gather up only those with at least half of their flesh looking good. We have accumulated fifteen sacks, half-full and feeling sodden.

At eleven-thirty, a dirty blue van rolls up and some villainous-looking men stumble out and ask if we are the apple people. The load-up is over quickly and we ask when the cider will be ready and will it be delivered? We get some strange looks and wonder if the deal to make cider in exchange for apples has been misunderstood.

'Yeah, yeah, you can have some cider, we'll let you know . . .'

Pear harvest

I have held the fattening pears that hang on the espaliered fruit trees along the wall, supporting them carefully from below to let the full weight fall into my hand. Very gently, I lift my hand to see if they want to break from their parent stalk, but they are hanging on firmly. Not ready, not ready.

But the monster pear tree in the garden is shedding pears and tempting wasps. They are small, butter-coloured pears with a rough texture and freckles and if they are picked at the critical moment, they are the most delicious pears in the world. A day too early and they are hard and crunchy, a day too late and their middle has turned to cotton wool. Judging by the fallen, we need to pick now. The tree is crowded out with them, but I can't reach any. There is no point hitting with a stick; no point in risking life on a wobbly ladder; we must wait for Michael to come home and make it a family exercise.

He joins in enthusiastically. Georgie climbs onto his shoulders and stretches up and then hands down her pears to me, waiting below with a basket. We reckon there might be a week's worth of harvesting to do, some each day until the cotton wool and the wasps take over.

For supper we eat pears.

Birthday rose

I asked for a rose and it was delivered with a caution: this rose will grow into an impenetrable thicket. Is this a selling point? For a fairy tale? It also will have a million small pink flowers. This sounded good and we thought carefully about where an impenetrable thicket would be useful, even when flowering. There was a dull end to the long, thin flower border and a riot of pink flowers would look ravishing from the kitchen window. We dug a hole and offered a heap of goat droppings and trod it in. 'Good luck, impenetrable thicket,' we said.

And now it is that autumn day when the impenetrable thicket needs to be hacked back to stop it raging out and taking possession of the whole garden. Its stems are like barbed wire twisted around steel cables. I am a warrior dressed in an armour of strong clothes and I make the first incursion by the privy, where I can crouch down under the briars and make an awkward attempt with a bow saw.

I also have long-handled loppers with a strong bite. It's easier if I lie down and lop upwards and twist the handles to stop the barbed briars falling on my face. I wriggle out and a gloved Michael and I start yanking these huge angry stems onto the lawn to be sawn into manageable lengths to be dragged to the bonfire. I discover delphiniums and lilies cowering under the impenetrable thicket, blighted by their overpowering neighbour.

We are brutal. We hack and lop and chop and saw and snip and cut and pull and drag. As far as the impenetrable thicket is concerned, it's nothing more than a haircut. In the spring, it will burst into leaf and by early June, it will be an impenetrable thicket of glorious pink roses.

Washing a carpet

I have a clever plan to sort out a carpet. The enormous and beautiful carpet that we found in a ruined state in our previous house has been spread handsomely in the drawing room, but it smells. It smells slightly of cat pee. I have hoovered it a thousand times. It has the benefit of being stretched out in a warm and sunny room. It has a log fire next to it. I have sprinkled products on it, like Remove Cat Pee, or Blossom

Room Fragrance, and even Baby Powder; the faint but pervasive smell of cat pee remains. I have an idea.

It's a beautiful week of weather, hot and dry. The front lawn has been meticulously mowed and is not covered in stray snips of grass.

I enlist a reluctant Michael to help me roll the carpet. This involves only a little bit of hoicking up the oak chest so the carpet can be pulled through, and the big sofa up a bit too, and the removal of a few chairs into the other room. We are ready to roll. It's a huge beast, three metres by two metres and it makes a large roll, reluctant to bend around the doorway, but with heaving and pushing and tugging and rolling it is outside and ready for washing.

I unroll it and bring the cold-water hose around to the front. The carpet drinks up the water like a camel and when it is utterly soaked, I bring out a big bowl of hot water with highly fragrant washing detergent and throw that on the carpet and begin to scrub with the yard broom. We have suds and bubbles and I have a flashing moment of doubt; was this a foolish plan?

I can't tell if there is dirt coming off the carpet, but once I have beaten in the soap, I begin rinsing and hosing and tread on the bubbly, sudsy carpet, worrying that I am making a sea of mud below.

Then I leave it for the afternoon sun to dry it, overnight to drain, over the next day to use the heat of the sun, over the next two days, over the whole week. I check it almost every hour to begin with and then lose heart as the carpet remains indefinitely soggy. The kind summer sun beats on, each day, and after a week, I think we are nearly there. Michael and I roll it, heave it up onto the warm concrete and leave it there for a few more days.

Finally, it is dragged back into the drawing room and I brush off any grassy bits and then hoover and hoover and hoover. It is almost dry, no lingering smell of dampness, but coming through we can detect a slight smell of . . . cat pee.

Mushrooms

A September day, properly damp but warm, spells mushrooms. Out in the big field we all spread out, scanning around us for specks of white. The first ones are hidden under leaves, but once we find one our blood is up, the hunt sharpens.

Morwenna is the keenest hunter of them all. She rushes like a hare up and down the contours of the field, calling out, 'A beauty, a beauty!' and comes proudly back with her mushroom treasure. She picks carefully, cutting the stems with a knife, then holding them flat on her palm with the fragrant, pinky-beige flesh uppermost. We discuss the darker-fleshed ones, older and probably holding small bugs; but we take those when the crop is small.

The target is to find enough for all of us to have a feast of mushrooms fried in butter, with some rashers of bacon and buttered toast.

Multi-tasking

Michael is multi-tasking on the phone. I hear him advising on a complex physics question to a student but this activity is above the waist. Below, Georgie is tugging at his trouser leg and his hands are occupied in fixing a bow in her hair. Good work.

Gathering nuts

How is it that we have what seems like rich pickings of hazelnuts from the hedge and a basket of walnuts gathered from the lane by Morval church and yet when we open the shells, everything inside is a dank, black, withered ex-nut? There must be a trick to this. Should I be drying the harvest in the bottom of the AGA cooker that I do not have? Should I leave them on the branches so that the squirrels get them first? Should they be laid out in the sunshine on the granite floor of the conservatory?

My *Guide to Gardening* says that the walnut tree grows to be a fine specimen. *Self Sufficiency* says that walnuts are good. In *Food for Free* I learn that pickled walnuts are easy to do, yuk, and that it is hard to get to hazelnuts at the right time – too early is too tasteless and too late has the kernel turned to dust. Where are the helpful 'How To Do' instructions?

We are not destined for nut success.

The hour of industry

Michael is running a school in our kitchen. His cleverest students are here in a huddle around the kitchen table and are being given extra lessons for an Oxbridge entrance. The school's science excellence is welling up and pouring out in our kitchen. It's early evening, before the animals get shut away, before our children need their supper, before Michael has to turn to other commitments. There are five boys, chewing on pencil ends, challenging and mocking each other, rising to Michael's questioning.

Georgie and Morwenna are fascinated. They creep up and hang onto Michael's arm, each pushing a proboscis into the action, smiling shyly at the big boys; they want to join in, it's their kitchen, too. Michael sees this and leaves the boys with a set problem and settles the girls on chairs close by and gives each their own problem.

For Georgie, he has a sheet of drawing paper and she has to join the dots he has drawn to make a crocodile. For Morwenna, some sums, times-table ones. They each settle quickly, swinging legs and humming little tunes of happiness. They are part of the big table of thinking and working, some long, spindly pubescent boys and some small, pink shiny girls.

I pull out my grant application for the cottages. The hour of industry has begun. We can do this forever.

Chapter 6

Hedgehogs

Jean has delivered four tiny hedgehogs in a box. She found the orphans in a nest beside a dead hedgehog and has given us detailed instructions for their care.

Before unpacking them, we have to make some rearrangements in the greenhouse. We empty it of mouldy tomato vines and put a wooden wine box upside-down and propped on stones to make a winter house and line it with hay. We tape up the draughts from the ill-fitting glass panes and bring in a greenhouse heater, which has a little lit flame and gives off a pathetic level of warmth. The baby hedgehogs are to be given cat food and water every day and this should build up their body weight to a point where they will either be able to hibernate successfully or keep on going until the spring.

Morwenna and Georgie are fascinated and bothered by the prickles and the fleas, but I don't think I can do anything about the fleas. It doesn't feel right to subject them to louse or flea powder, when they are probably suffering from stress. They have tiny, bright, black buttons as eyes and snuffling soft noses, and curl up quickly into a fierce ball when we appear. The cat food is eaten, the water spilt and replaced, and little piles of hedgehog mess appear. Today I see a glimpse of some scurrying feet disappearing into the wine box.

The right apples

I suppose we were foolish in planting the right apples, choosing old Cornish varieties that had names like Snout, simply because we were upholding a Cornish tradition. You would think that having once had a whole orchard of revolting cider apples we would be more hedonistic, longing for honeyed loveliness or crisp, refreshing juiciness.

In the vegetable garden, lining the fence we have a row of apple trees with too many undeserving, unyielding, unappetising apples and too few delicious old friends such as James Grieve, which are beloved by the wasps and me.

We prune them all in textbook style. No branch is allowed to flirt or linger in the centre of the tree, each branch must thrust outwards and bear fruit buds. We snip at the right length. And as we do this, I hiss at the dullards. This Cornish Gillyflower has had only two skinny, tasteless apples this year. What we should do is some grafting. Cut down the vigorous main stems and bind onto them cut spurs of some properly delicious Beauty of Bath, some Pippins, some Worcester Pearmain. I have a plan.

More apples

The cider from last year is never to be repeated. It was thin and yellow and sour and smelled musty. So we still have an apple issue. I have loaded up several barrows and thrown them into the field for the goats and some more into the pig trough. I am alarmed to find the pigs giggling in the corner, totally off their feet in a helpless pool of mirth.

'Pigs,' I say, 'what is going on?'

They don't really need to tell me, because the smell of booze emanating from the trough makes it clear that the pigs are drunk, and I have been a poor parent in supplying them with fermenting apples.

Halloween lanterns

It's Halloween and I have not been paying attention. I haven't bought pumpkins for carving and when I remembered this morning, the vegetable shop had sold out. I tell the girls it's no problem, we can make the weirdest and best jack-o'-lanterns.

I'm not sure if I have sold this well enough, but while they are busy, I rush out to the vegetable garden to have a fresh new look at our courgette plants and to consider the rejected marrows as possible salvation. It's not impossible. There are two marrows whose noses are long and thin, and they bulge out into fatness at one end. There are two chubby, banana-shaped ones. I harvest these and stagger back with my spoils.

The girls look doubtfully at the collection, but I draw sketches of how they might look and offer to get out cutting boards and knives, and start on the scraping and hollowing. It is quite exciting and we all get busy scratching out the unwanted flesh. I briefly consider whether we should use the marrow flesh and then remember that we hadn't wanted the marrows in the first place, which is why they are lying about in a rejected and ignored state at the end of October.

We manage gappy teeth and slitty eyes and a menacing smile on the chubby marrows and think hard about the long-nose ones, before deciding maybe two marrow lanterns is just fine. In a moment of pure theatre, we light the candles inside them and turn out the lights.

AAAARRRGGGGHHHH! We have Halloween monsters of a very different, greenish kind. You don't HAVE to have pumpkins.

Collecting the cured ham

I have driven miles into the remote, distant Cornish countryside to collect the hams. It feels like a pilgrimage, a medieval journey to a craftsman, far from home. I took half a pig up here some weeks ago, carrying joints of freshly slaughtered pig hocks and legs and sides to be turned into bacon and hams. The farmer had met me in the drive and taken me to his sharp-smelling workshop, where he had casks for curing the meat in saltpetre and other ingredients for his alchemy. There were encouraging joints hanging there waiting for collection.

Now it was my turn. He had wrapped the joints in heavy-duty sealed plastic bags, but I could see how fine they looked, brownish, the right shape, and the bacon already sliced for me looked fantastic. Like all good harvesters, I felt thrilled and proud to be bringing this home to feed us during the winter festivities.

And now the long drive home in the gloomy twilight, down lanes with high Cornish hedges and magnificent, bare-branched winter trees; the thin, wintery light fading, and my family at home waiting for their supper.

Goats to the abbatoir

This is a morning of utter treachery. Two bouncing and lovely goat kids have to go to the abbatoir to be turned into meat. They have been our friends and playthings and now we are disposing of them. We could sell them, our vegetarian or townie friends would cry, but there is no market for billy goat kids, castrated or rampant. This is farming.

Today's arrangement is that Michael is taking them to the abbatoir before school. He is dressed in a tweed jacket and proper school trousers, and he has stuffed himself into a boiler suit over the top. His school shoes are in his lesson box on the front seat of the van. I have helped him lure the greedy little goats into the back of the van with bowls of extra food and they have nipped in like unsuspecting angels.

I know what happens next. He lurches up the hill into Liskeard and the goats slither about in the back of the van, bumping into the sides and complaining, but when the hill is finished and the road is smooth, they explore their little compartment and chat to Michael about it. He replies in soothing words and it is all very friendly. They come forward and start to nuzzle the back of his neck. They nibble at the collar of the boiler suit. They breathe hot whiskery gusts down his ears. They offer to climb over the seats and join him on the front seat with the shoes and schoolbooks, but he manages to shoo them off with a snort and shake of the head.

They are in friendly chat all the way to Tideford and I suppose Michael's heart is churning because he feels bad. As it is early in the morning, no other creatures will yet be here, crying out or frightened, and the little goats come out hesitantly and are led into a shed by the very jolly man whose job is to be executioner.

Michael, do you give a goodbye rub onto a soft nose or behind an ear, or do you turn and hurry away?

I must return in a week when the carcasses have been inspected and passed and the meat cut up into chops and legs and shoulders. We tell our friends that they are eating lamb. The children will not touch it.

Juniper

Juniper is the lowest in the pecking order in our herd of sixteen goats and much despised by Yamaha, the queen of the flock. Yamaha is sleek and white, tall, haughty and with a wicked, knowing yellow eye; she gives the orders to the rest and they follow. She has first food, is led out of the stables first and that is that.

Juniper was donated, an unwanted orphan, by someone who had heard that we kept goats, and she arrived as a skinny teenager who was afraid of life itself. We argued about a suitable name – obviously she needed to be named as exotically as her appearance. Something spicy and foreign: Cinnamon? Anise? Juniper.

She is mid-brown with fur that is as soft as cashmere, she has long floppy ears pointed like mangoes and of a sleeker, darker shade of brown, she has a roman nose and black eyes. She should be part of a harem in the Mediterranean, but instead she is here in Cornwall in the rain and Yamaha hates her. I witness bursts of bullying and am appalled by Yamaha being so racist against the only brown and beautiful girl in the herd. Yamaha is clever, Juniper is hopelessly stupid.

Today I try to forgive her for being so stupid. She is in season, which I know and she doesn't. She is bleating and flicking her small brown tail about and I know that I will have to take her to the billy goat within twenty-four hours. It is 4 p.m. and we are supposed to be going to the theatre tonight, plus collecting Phee to babysit, plus feeding children and getting ourselves tidy and fed. I ring Sylvia to ask if I can bring Juniper around and then grab some essential equipment (string for leading, boiler suit) and a distracted husband.

The normal routine is to lure the goats up the plank into the back of the van and off for a very noisy, bleating journey to the billy. We observe a tense introduction, a few minutes of amorous foreplay, a three-second act of love, and then repeat the entire process in reverse order with the atrocious smell of billy goat reeking in the van. This should and could only take an hour and a half, most of which is spent driving through the smallest back lanes of Cornwall.

For a first mating experience, I often let Yamaha come along for the ride, thinking she will reassure and tell the dark secrets of goat love to the young bride, but this won't suit Juniper. We enter the goat shed and put the string lead around her collar and attempt to lead her to the van, but she won't budge. We try tugging, Michael at the front end and me giving her a good shove from the back, but that is useless and the food bribe isn't working either.

Time is ticking on. It has to be tonight. We tug and shove and tug and shove. In despair, I go to the food barn and wheel back the rusty old wheelbarrow and Michael heaves Juniper into it with a great lunge. She is so astonished that she sits there blinking and is trundled along to the van, up the plank, and is tipped in.

We drive off to Sylvia and her billy goat and when we arrive at the dark wuthering-heights establishment that is run by Sylvia and Norman, there is more idiotic hopelessness from Juniper. The smell of billy goat has no impact on her at all; she shows no eager wish to leap out and investigate. She sits curled up near the driver's seat and won't be tugged out. Sylvia remarks that she has never seen a goat look less ready for the billy.

Is there a wheelbarrow?

Juniper is pushed into the billy's shed in Sylvia's wheelbarrow and tipped out. He is on fire with desire and snickers at a disinterested Juniper. The billy breathes some seductive sexy message into her ear and suddenly she gets the idea. It's over in moments. She is mated and returned surprised; all by wheelbarrow.

Carol singing

The weather is just right, chilly but with clear skies. We hastily munch a very good tea of drop scones and head out, bedecked with hats, gloves, double socks and wellingtons.

The carol singers are myself, ferociously keen and definitely the leader on this one, Michael who is reluctant and pessimistic, fearing imminent snowfall, wild bears, heart-failure along the way, and missing the News. Both girls have been revved up by me into thinking this could be a thrilling adventure, seeing it is dark and there are torches. Georgie is in full voice with the only carol she knows so far: 'Jingle Bells'. Morwenna is tugging at my arm to get going.

There are three houses within walking distance and we surge down the drive with a skip and a hop and reach the little cottage on the road. The gate creaks open and the overhead security light bursts a blaze of too bright light on us, intimidating and silencing us all. We have a choice of two doors, one looking cobwebby and unused and the other with some bowls of cat food, so we cluster round and I have Georgie firmly by one hand steering her away from stomping in the cat food.

We start by singing, 'We wish you a merry Christmas, we

wish you a merry Christmas.' We sound a bit feeble, so we launch into the figgy pudding verse with some gusto and suppressed giggles. We can hear shuffling inside the door. It flings open and our surprised lady neighbour registers disbelief, quickly adjusted to neighbourly delight, and insists on us coming in. This flusters us, as we are dressed for the North Pole. Do we strip off? Too complicated, so we troop through to her living room and stand awkwardly by the Christmas tree.

Morwenna wants to sing 'Away in a Manger', so we do one verse and I nudge Morwenna to indicate the end, as I am in danger of serious overheating. We exchange a few Christmassy words and file out again into the frosty night. Hurrah, we are a polished performance now – we have got the system sorted.

We decide next to target our beloved Ernest and May across the valley. It's a single-file walk along a road not really wide enough for us and traffic, so we dangle the torches and lantern assertively.

We creak up the little pathway, jostling lanterns and torches and stand outside the door to start up again with 'We Wish you a Merry Christmas'. Morwenna and I are good on this. The door opens and Ernest and May come up and beckon us into their little porch. This is much better: cold with starry night beaming in. May asks Georgie if she knows 'Away in a Manger', but Georgie, carried away by the romance of the night, says firmly that she can sing 'Twinkle, Twinkle Little Star' and launches into a solo. It's wavery and squeaky and rather lovely, and we head back down the path glowing with success and praise.

We have almost run out of reachable neighbours now, but risk all by passing our own farm and carrying on up towards Jack's. It's a long old walk and we sing as we go. There are no cows out to greet us, no dogs, no sound at all, not even the muffle of faraway traffic, until our boots shuffle on the concrete drive. The steam from the silage gives out an atmospheric, misty feel and we start up our shrill little carolling.

It's so quiet, we sound shockingly loud. We knock on the back door, but no one comes out, no dogs bark. We blast on for what seems a miserable age and then slink away, defeated suddenly.

Bough gathering

The Christmas family gathering is coming together and the first arrivals are being sent out to gather winter boughs with which to 'Deck the Halls'. My parents and small children are dressed up against a cold wind and we set out with hessian sacks and a supply of secateurs. We traipse down the farm drive and cross into the forest. The children are excited and eager to get picking and keep pointing out unsuitable muddy branches, which we agree to cut but I chuck when I can.

Soon enough, there are some fat pieces of fir discarded by the forester and we cut handsome bunches and stuff them into the sack. Morwenna knows where the holly is and leads her cousins at a fast pace ahead of us old ones, and they race away full of excited chatter, splashing each other in puddles on the path.

At the holly place we gather and pause, wondering how to bring down the swathes of berried branches that are out of reach. We try the adult jump, twice, and the small child hoist, which leads to a collapse in the bushes below. My less impulsive father finds a long, hooked stick to bring branches low enough for children to cut and gather. We cut and cut, needing fir, holly and long curls of ivy in endless spaces in the house. Our haul triggers a bout of carol singing and we set off loudly with 'The Holly and the Ivy'.

Hedgehog party

We nip out in the moonlight to see if there is any hedgehog action. Quiet, creep, shhh and there in the greenhouse, lit up by a waxing moon, is a hedgehog party. Four fat little hedgehogs are not hibernating at all, they are munching cat food.

Christmas stockings

At 4:30 it's dark and the children are weaving and jostling in every room. Clearly they need supper. I have made a vat of macaroni cheese and settle them all with great bowlfuls. With their mouths busy eating, a heavenly peace falls upon the kitchen.

Next it's time to choose a Christmas stocking and put a label on it, so we can nail them all onto the kitchen beam. 'One for everyone, even grown-ups,' I tell them. I have a stack of stockings in a mix of reds, Christmassy prints, red bows, white fluffy trim, festive ducks, and velvet. Some of them are ancient and have served many Christmases already, and some are new in honour of this large family gathering of twenty.

Soon I have willing scribes at the roll of sticky labels, wanting to know which names to write and how to spell them. 'Which stocking holds the most?'

Michael fetches his small hammer and tacks and we start up a great line of dangling stockings knocked into the big beam with names visible, so Father Christmas won't get confused. Our kitchen has never looked this festive before. The windows and the big mirror are covered with cut-out snowflakes, the wooden pan rack suspended from the ceiling is hosting boughs of holly, the centre of the table is a cornucopia of berries, baubles, little birds, golden pears, holly, fir branches and candles; the mantelpiece has a cluster of medieval angels and there is Christmas music playing.

Early daffodils

On the top of the bank along the drive there are strong shoots pushing up through the grass, and each daffodil shoot is tipped with a tough point. It's deep winter, but these little flowers are warriors.

They came from the old daffodil field down in Millendreath, where they had reverted to growing wild over the years. They had then grown strong when we planted them on the banks here, nourished with fertiliser and unencumbered by brambles. Fighting off winter, their pale blooms stand up to frost and gales.

Morwenna picks me a bunch and posts them into the Quimper jug, and swings her legs as she draws a spikey picture of them.

Green food

A nourishing meal wrought by my careful hands and with home-grown ingredients: a leek quiche. Delicious, but Georgie stares glumly at it and will not be persuaded. 'It looks green,' she says. 'I don't eat frog.'

Harold lost

Rosetta is in the field and has started bellowing. We rush out to see what is causing her so much grief, and her newborn Childe Harold is not at her side. We obligingly start to search for him on her behalf. The field is large, the new grass is long, and there is no sight of him as we scan across the heads of grass. We scatter and each look through a quarter of the field, and meet up with sad Rosetta with still no sighting of Childe Harold.

I get a horrid feeling that maybe he has strayed into the river and slipped, fallen, or drowned. I have an image of a crumpled body lying in the river weed. So we scrutinise the riverbank, a careful search all the way down and find nothing. Rosetta's bellows are painful to hear and we can't believe that a newborn would not respond to such a summons. We look in the barns. We start to pace through the grass again in a solemn line.

He is there in the grass, completely asleep, lying in an almost invisible ball of soft brown fur. We squat down at his side and rub him awake. He sighs and blinks. Now awake, he hears his mother and struggles to his feet and makes his sleepy way over to her. She is all forgiveness and nuzzles him, before nudging him towards her udder where he pulls a long lazy drink. She looks out across the valley, patient and content, with a flick of her tail.

Bra for Rosetta

Every morning when Michael goes out to milk Rosetta, I worry about the split on the side of her teat. Childe Harold suckles his mother ferociously, I know, because he almost rips the skin off my finger if I offer it to him. The old lore for managing a house cow is that we milk her first and take what we need and then Childe Harold can join her and suck for the rest of the day. Then they are to sleep separately. It seems to work, but I fuss over this sore that she has. Childe Harold's sucking is making it worse. Maybe it doesn't hurt her, but when I find flies sitting in the split, I immediately imagine they are going to lay eggs and that I cannot bear. We have to heal her.

I have tried soothing in udder cream, but Childe Harold just licks it away, and worse still I have seen Rosetta contorting her body and licking it herself. I have tried putting a plaster on it, rather thrilled to find a use for the huge size in the mixed box of plasters, but of course that came off almost quicker than from a sore heel in the swimming pool. I have even tried a plastic skin paint. Needing to think of something that would really work, I put myself in her situation. The only thing to do if it were me would be to put my sore nipple away very firmly in a good strong bra.

Of course! Rosetta needs a bra. I spend a small time scratching away some good designs on a bit of paper, but realise I have measurement problems, so I pull on my boots and with a tape measure in hand, I set off to find Rosetta in the field. She is wary of my plans. I throw the tape over her back enough to dangle down to udder level and realise the task is longer and bigger than my tape measure. So I measure from the start of her udder to the middle and guess the sort of cup size she might require, taking it as given that the tying strap will have to be considerable. She flicks her tail to show that my interfering is annoying.

Pleased with the project thus far, I then rifle through my fabric boxes to find the best material for the job. Should it be flowery cotton? Something tougher? I find a huge piece of net curtain: perfect. I get out the sewing machine and make an enormous bag with darts shaping a spectacular cup size and then attach mighty straps to tie over her back.

Trying to dress Rosetta is far less successful. I abandon doing it on my own and enlist Morwenna and Georgie's help while I try to tie it, but Rosetta gets frisky and annoyed and skips off. Securing the cow bra is going to be a two-adult job.

Bra fitting

It's a full team effort this morning to get on the bra. The children are excited and carry the great frill and folds of net, trailing the long ties behind them.

Rosetta is milked by Michael, who tethers her to the railing, drags the stool to her hindquarters, pulls in the big stainless-steel pail under her udder and starts to squeeze. He says it's the best moment of his day, so we leave him to it and the girls and I go and play with her calf, who is hungry for his breakfast and licks and slobbers our hands. He bellows for Rosetta, who answers him.

Michael calls out that he has taken what he needs and so we clang open the barn door and let Childe Harold bound over to his mother and start sucking. He pushes violently into her udder, but she simply chews the cud patiently and unperturbed. When he is done, he breaks free and nuzzles her. This is our moment for the bra.

I slurp udder cream onto each of the teats, then Morwenna brings the bra over and I administer a proper Marks and Spencers' fitting service, tucking in all the teats and making sure it's a snug fit. The ties are fixed firmly in a flamboyant bow on her back where I am sure she cannot reach, and she is let free to do a day's wandering and grazing in the field with Childe Harold bouncing beside her.

I can't resist a snoop mid-morning and stand out on the ha-ha to see where she is.

She is not immediately visible, so I jump down to look for her. She is down in the sloping corner of the far field, through the gate, where the bluebells are thick. She has her calf with her and there is no bra. 'Rosetta, where is the bra?' I trudge back and make a circuit of the field, but don't see it anywhere.

After school we all go on a bra hunt, fanning out in a line like forensic experts. We find it trampled brown with mud. I am not discouraged; it comes in for a wash and will be fitted again tomorrow.

Grooming goats

We all love this job. It's a Saturday morning sort of job, where time can be spilled carelessly. The concrete path has been hosed and scraped clean and the slush pushed into the slurry heap. The animals are fed, the ducks and geese are in the river, and the hens are happily chattering in the yard. We are in the two goat sheds, and the separating wall of the manger is full of hay.

Each of us has a coarse brush and we are grooming our goats, giving much happiness. The brush picks up strong, coarse white hairs in thousands and we peel them off in clumps and chuck them on the ground. Each creature is sleek and glossy and astonishingly white or cashmere brown. Brush noses, faces, behind ears and pay attention to that absurd little goatee beard.

A routine look at hooves is easy, but cutting hooves is not a Saturday morning job; it is arduous and smelly and must be planned. This morning we have patting and smoothing and loving time and then we lead the herd off into the field and stand about watching them. They nibble politely at our feet for a while and then with courtesy attended to, they follow Yamaha towards an out-of-reach forbidden corner.

Murder?

Our morose GP visited a sick Georgie and remarked in a tentative sort of way that it was surprising that we didn't have ghosts in the house, as the last time he came to this house and indeed this very bedroom, the incumbent had been old, feeble and ranting about his past misdeeds. What misdeeds?

'Well,' he paused, to justify revealing confidential information. 'Maybe the old man was experiencing a bad dose of guilt over the mysterious circumstances in which his brother had been found dead.' No verdict, no crime, no explanation.

We thought about this, but agreed we had never felt any unexplained shivers or clammy moments in the night.

A while later, a man came round to give costings on the replacement of our roof. 'Ha, this is the place where there were those murders,' he said. I lured him inside and made him coffee and tried to get more of a story from him. 'Everyone knows,' he said, 'two brothers and a sister lived here in jealousy and bitterness, and they say the brothers drove the sister to death of a heart attack by circling round and round her in the field in their tractors.' Later, one of the brothers was found dead in the river. The third brother had lived out his days here in the house, alone.

All this gossip and rumour. When we, a young couple with a new baby, first moved in, Jack said that it was right and good that at last the farm should have young people and some happiness.

No one has more to tell, so we are going on in happiness.

Spilt milk

Michael comes storming into the kitchen cursing. Ha! I can see in a flash what the problem is. Again. He is carrying the goat milking bucket which is empty and muddy on the rim and his trouser leg is soaked. Not quick enough in your reflexes? I enquire, or did Yamaha kick even through the strongman hand grip? He snarls.

Chapter 7

The privy

In the garden there is a privy. It no longer has a long drop or the pair of wooden seats, and we gave it a proper door, thinking it would be a good place for storing lawn mowers, until we found we couldn't reach the door past the impenetrable thicket of the rose. And so it languished unused, dank, collecting cobwebs and dusty, dried leaves and a straggle of etiolated nettles.

But Georgie is having her fifth birthday and in great secrecy, we are bringing the privy back to life. Carpentry, measuring, paint, upholstery, furniture, plants. We whisper conspiratorially. Can Morwenna be trusted not to squeal?

A friend has come to help and has pulled off the metal flap that formed a window, instead putting in a glazed window and frame. We fix a ladder to a shelf bed which we have made and a foam mattress covered in a bright design. Underneath, Morwenna has dragged in the little wooden table and chairs, and a small cooker made from a bedside table with a tacked-on backplate, aluminium knobs to twiddle and three painted hob circles. There's a small rug on the floor and white paint on the walls.

Morwenna is flying in and out of the house bringing in life's necessities: tea sets and the saucepan from our camping set, while I have dug a flower bed outside the door and planted up some primroses. 'Can we sleep the night there?'

asks Morwenna and I say, 'Of course you can.' We are so excited by our gift.

Morwenna drags the birthday girl out of bed and out of the door; we hurry after them. Georgie thinks she is heading off to the animal sheds and is astonished to be swerved into the privy. Morwenna pushes her in, follows, and shuts the door. There is very loud squeaking to be heard.

The haystack

The haystack ebbs and flows. When depleted it is a sorry thing, the last bottom rows of blocks seem the mouldiest, scrappiest and most dingy; worse still, trying to heave them up is the hardest. They have become embedded into the dry dust below, their baler twine string mostly broken, and we lever the loose flanks into the wheelbarrow for the goats, who look sadly at this offering and turn away. It's scarcely bedding quality, giving off clouds of dry mildew when it's broken and scattered.

So we search the pages of the *Cornish Times* for farming equipment and find an advert for horse hay, expensive but at least in physically manageable bales of the old rectangular kind, not the new-fangled rounds held high by tractor prongs.

But in times of plenty, our hay is piled twelve layers high in the barn, stacked like an Egyptian pyramid by us, the slaves. Bringing down the bales for greedy cow and goats means climbing to the top and jostling a few down and breathing in deep the smell of hay, and there are few other smells so stirring of the soul. Hay work means wellies and long trousers against the flying fragments that sneak into any loose cuff, so we might as well bring down six or seven to the lowest level for easy tipping into the barrow on less well-dressed days.

Michael grabs a bale in each hand, I just take one and use my thigh to help carry the load; the girls push the barrow. Sometimes we pile up the barrow four high and then an attendant moll, Morwenna, is required to steady the journey through the yard. Heaving the bale up and into the manger is a muscle job. Up onto the shelf, then a great heave to the lip of the long manger, seven-foot high. The haybale drops down and the goats leap onto the shelf to forage for morsels of interest.

The haystack is a secret city. It shelters cats basking in a sunny spot. It houses random collections of the girls' toys, laid out for a picnic; it has the remains of a camp with an old frying pan and a cushion; it probably contains families of mice. It has a view, down into our garden with lawns and the island of pear and apple trees. Really, one of the fine things to do is sit on the haystack and muse on life. Is there time to muse on life?

Goat skins

I am fed up with my ridiculous projects. Why am I curing two goatskins on the conservatory door? What possible use will they have? Is it to prove I am Queen of Self Sufficiency? Or because I can't bear to chuck out Mango and Spotty's luxurious, thick pelts, in shades of nutmeg brown.

Whatever, I have soaked the skins in formaldehyde, which has given me multiple and unbearable headaches. This has taken a couple of weeks and now I have stretched the skins onto the door, hammering carpet tacks into edges that defy all grip as they slime away from my fingers. Then I slop lanolin mixture into the skins, rubbing it in as if it's a spa treatment, repeating the massage as the skins dry out a little and the stench of lanolin mixes with the ghostly echoes of the formaldehyde.

Michael makes a huge performance of scrunching up his face and hurrying past, gasping as he goes and telling the children to run past without looking. None of this bodes well. I tell myself it is a survival skill, good use of resources, a defiance in the face of waste, exemplary.

Morwenna and the goat skins

It was bad enough curing those skins, but now that the best one is beside Morwenna's bed and the smell has faded, I am struggling with her attachment to it. She has developed a little ritual before she gets into bed; she stands on the goatskin and puts her feet together and twizzles them back and forth and then digs her toes into the fur and says, 'Goodnight, Mango.'

Michael and I have to talk about this. She has always coped well with the matter-of-fact business of life and death on the farm, helping with plucking and watching as I gut out the carcases of chickens and ducks, joining us in our fury at hatchlings being taken by predatory birds, nodding wisely as I explain about Mango and Spotty being taken to the abbatoir. But now I am worried that we might have got something wrong. 'Goodnight, Mango' is bothering me.

The train

We have a train that follows our river boundary. Our train is only a single carriage for most of the time on a branch line and we love it. We wave to it from the edge of the ha-ha and holidaymakers on their way to Looe wave back at us. The children as toddlers would stand still and wave, and as older and wild farm kids, they run helter-skelter from wherever they are playing to wave. Waving from the vegetable garden, waving from the meadow, waving while leading goats, waving while pursuing ducks in the river, waving while paddling by the spring or in the watercress bed.

Our train rumbles on the rails as it comes up or down the single-track line once an hour; inside the house it works as an alarm clock, making its first journey at seven in the morning, when we should all be up and busy. It puts itself to bed well before we do.

During winter afternoons when the animal time is at twilight, the train clatters past like a Chinese dragon, lit yellow, shining bright and empty of holiday visitors. A late walk along the ridge of the sloping fields gives a view down along the valley and the shining dragon's route up or down can be seen for half a mile before the railway track twists around the corner.

And when the wind is coming from the north, the sound of the main-line trains can be heard, roaring through the night. Trains are our metronome.

Michael's choice

Michael returns late from an after-school meeting with the local councillors. He is cordially invited to be a member of a key group who will have access to the bunker prepared for a nuclear emergency. They think he will be useful as a scientist.

Oh? And? 'I said, no thanks,' he informs me. 'I said I would rather die with my family.'

Liberation from the railway gate

A notice has arrived on the wooden gate that is the barrier separating the branch railway line from our road. It is a sort of planning notice, announcing that the gate will shortly be removed. We receive this news dumbfounded.

It is exactly what we want, but we hardly dared believe it would ever happen. For several years we have endured in rain or storm, or urgency to be home, the process of having to stop the car, get out, open the gate, get into the car, drive over the railway line, get out and shut the gate and then whizz home, probably wet.

Now, ahead of us lie years of easy rail crossings. The train drivers will see our car bumbling down the lane and will hoot to warn us that they are close, and on good days we will stop and wait and wave. On late-for-school days, I will make a sprint and dash over ahead of the train and the driver will hoot anyway and wave his hands and shake his head, telling me I am a lunatic.

Elephant Fayre

It's festival time. Morwenna and I spend an hour in the Japanese Noh Play tent up in the higher grounds at Port Eliot to rest. She has been careering around with other feral eight-year-olds to the point of exhaustion, while I have been running the Lost Children's tent, increasingly alarmed by the fact that we have had one lost boy with us for more than twelve hours. The spaced-out parents appear and claim he has been missing for less than an hour. Michael has taken Georgie home for a bath and sleep, promising to return in the morning with a chilled box of milk and raspberries.

The Noh play has strange drumming and slow-moving dancers in pastel silk kimonos and we are drummed into a trance.

We emerge to twilight and the festival site spread out below us. A thousand tents of every colour and a hundred campfires are burning orange and sending spirals of smoke. It could have been a medieval pageant, an ancient battle camp. Morwenna asks, 'Are we in heaven?'

A tortoise

Last week we bought a tortoise from the pet shop. The vendor was cagey and weird about it and we nearly had to present passports. This tortoise has come from far away, is an endangered species and in need of responsible care, and I have agreed airily that we can provide all the necessary responsible parenting.

At home we are delighted with him or her and have set a cage in the conservatory and offered hay and a box and lettuce and some fragments of apple and the creature has seemed content and munched its way through these things. I realise I know nothing about tortoise care; how much salad it needs or if lettuce and apple are adequately nutritious.

A trip to the library to investigate tortoise requirements is scheduled for soon. But today is a day of sunshine and we decide to let him outside to forage for his own choice of grasses, moss, weed and other delights in the garden. He moves surprisingly hastily from where we release him and is seen striding off towards the end of the garden. He can't go anywhere as there are walls, a ha-ha and a steep grassy bank, so I am not worried – he has half an acre of his own freedom.

Then we can't find him. Surely, he must be visible somewhere? We look under bushes, round dank corners, in the long grass, in the old privy, but nothing. I call off the search and tell the girls that we can resume the hunt tomorrow, when he may have decided to come munching in a more obvious place, but I am feeling rattled. How can he have disappeared? Do tortoises climb walls? He wouldn't leap off the ha-ha voluntarily? With a sinking feeling, I realise I have been a reckless parent.

Open and shut the barn door

As I slam the kitchen door, ready to stomp off in high fury, struggling with my wellies, I can hear Michael shout out after me, 'Be careful how you open the barn door!' Aargh. Aargh. I have lived on this damned farm for years now, opening and shutting the barn door on a daily basis. Why must Michael persist in telling me how to do it? I know. I know. I know. I grit my teeth as I stumble down the drive, cursing him under my breath.

I let the gate into the yard slam deliberately, clanging its metal spike onto the granite post. I am at the barn door. Watch me, barn door. I know how to open you. I lift the metal handle and work it up and down to slide it across and then the door is free, but watch me, I am restraining the heavy old door, not letting it jerk back with its great weight and break the hinge. Watch, I let it go back in a controlled way, onto the string that holds it back.

See, I can do it. Me. I know. Clever me. I can do it and I DON'T NEED MICHAEL TO TELL ME HOW.

Blackberry time

There is a patch by the cottage wall where the brambles reign supreme and savage; we keep them there loyally for the fine fruit they produce. There are thick stems that surge upwards for grown-ups with hooked sticks to reach and there are stems that bear low fruit for small fingers.

We send the girls out on their own with bowls and warnings. Brambles have prickles, take care. We watch from the conservatory as the small one shows the bigger one where there is an unreachable berry, and with their heads down looking at their filling bowls, and as the small one spills all hers in the grass and they both stoop to pick them up.

When they come back, they have deeply purple fingers and a steady set of purple streaks down their fronts and glorious purple smiles.

Conversion of barns

I have a plan, and Michael thinks it's a good one, to replace my perilous occupation of part-time mother, farmer and my little extra business of running a summer language school for foreign teenagers. There is a government directive to get people like me into gainful entrepreneurship. I need to apply and attend courses in how to run a business and make and follow a business plan with cash-flow diagrams, and if my business plan is approved, I will get a grant. It is and I do.

I have to reapply for planning permission to develop two of the barns into holiday cottages, which should be easy now that the railway crossing has lost its annoying gates. I wrangle with the planning office. I have the gift of an architect in Geoff, who rumbles down from Liverpool to look things over and comes back with two inspired and detailed plans for a cottage from the tractor shed and overhead hay store, to be called Owl Cottage, and a single-storey cottage from the milking parlour, to be Kettle Cottage.

I have to find a builder and this is hard. Everyone is busy. I finally find one, who is inevitably going to be trouble. He has been bankrupt, but sells himself well and says he loves the drawings; he will follow them meticulously and there is hardly any need for the architect to come and check. I am a naïve fool and I believe him.

He builds and we watch, fascinated. I follow the plans and ask why this and that and annoy the builder, who says I am not to worry my pretty little head, so then I am annoyed. I ask his men and they give me worrying answers. Geoff comes down from Liverpool and explodes. They have done this and that completely wrong and I am to sack him.

I am between a rock and a hard place. Do I sack him? Where would the next builder come from? Like a proper businesswoman, I now have sleepless nights.

Burying a goat

My teeth are clenched and I am in full Worst Weather gear. And I hate Michael. It's been a very bad morning. Inside the goat shed we found one dead goat, and everything is my fault. She was unwell yesterday, off her food, and now she is dead. I should have and could have done something and the punishment for my failure is Michael's grim and silent accusation. Right now, we are doing Goat Burying.

Michael carries the long-handled Cornish shovel and thrusts a second one at me and indicates by grunt alone that this place on the far side of the hedge, near to the house and in the fruit garden, is to be The Grave. The girls are pressed up against the windowpane to oversee or sympathise; maybe they are holding a remote funeral service, but their presence is somehow comforting.

The ground here is terrible, a thick clay, but at least it yields to the slice of the shovel in a satisfying way.

It's hard work. I am wiping rain off my face and can't stop the trickle down my neck, and the handle of the shovel is slippery. We started off well, brisk slices around an imaginary corpse-shaped line, but under the top layer of clay there is resistant compacted hard soil and rock. Michael shows no sign of relenting or even commenting. I keep digging, pushing the edge of the shovel down with my boot and levering up whatever pitiful clod I can muster.

The rain is making a pool in our grave and it's harder to see progress. I ask Michael how deep we have to go, hoping for reprieve, but he gives me a look and I shut up. I haven't looked up for a while, but I know the girls are still watching.

The pile of soil is growing large and it's quite hard to flick the stuff off the shovel. I have taken up a position inside the grave to get a better lever. Michael jostles in beside me and we are bumping into each other, working independently. The grave is deep enough, I keep thinking, but Michael keeps on with a ferocious look. He has thrown off his coat and is in soaking shirtsleeves. I give up and sit on the edge of the hole we have made and stare miserably into it. 'Isn't this enough yet? Isn't it? Shall we get the body and see if she fits?'

Surprisingly Michael agrees and we trudge off in the rain and get out a wheelbarrow and in the other goat shed we find the rest of the herd sheltering from the rain, puzzled as to why their shed is shut, the shed with the body. We leave the wheelbarrow at the edge of the step and tip the body into the barrow and hurry off before we have too much fuss and enquiry from the rest of the herd.

We lower her into the grave, bend her legs with difficulty to fit, and step back to consider. Michael might possibly have had enough himself, because he reaches for the shovel and starts to fill in. Earth slapped onto white fur. It's backbreaking – shovel, shovel, shovel – and when we seem to be nearly covering her, I find I can't bear to step on the earth to compact it. So we load more and more on, making a proper sexton's mound.

In the conservatory, I lob off my boots and stagger into the washhouse and sluice the mud off my hands and face. I want a shower, a cup of tea, to cry, and to run away, but I go in to think about cooking lunch. I still hate Michael, superior and silent.

Goat bottles

Opening the paper sack releases an overwhelming smell, half nice and vanilla but cloying, and it forces you to breathe deeply outside the sack before rummaging inside for a measure of powder. Breathing it in would make you gag. It's infant formula for goats.

We have resorted to this as we now have two orphan kids and a mother who has died in childbirth; we have

Plum with twin kids; she hasn't enough milk to share with them, but we all know breast-feeding is best; three times a day formula and its hassle, hygiene and time is not my preferred option. The girls are devoted, though, and will willingly sit perched on the manger with a small kid nuzzling at a bottle full of vanilla froth.

We have been through several systems. Ordinary baby bottles have teats that are the wrong shape: too nubby and not long and droopy. Cornwall Farmers have lamb teats in red rubber with a small hole at the end, through which no formula will pass. When I enlarge the hole, it roars out soaking the kid and choking it. A further trip to Cornwall Farmers for their total supply of five, and I have prised a smaller hole with a tapestry needle, which looks better.

The formula instructions are hard to read through the sellotape with which I repaired the rip I made trying to open it. I think it says, 'water temperature blood heat', but I'm really guessing the quantity. It needs whisking laboriously, firmly, endlessly to avoid blobs of unmixed formula that are going to get stuck in the tapestry hole. It's smelling like synthetic ice cream. Georgie and I puzzle over suitable bottles. The only bottle size we have plenty of is wine and so we funnel the formula into two rinsed and warmed bottles, once holding Good Ordinary Claret, and fumble the lamb teats over the top. It's a snug fit, promising.

In the goat shed, we each have a small kid held between our knees, chilled, pitiful, lacking in mother love, and we try slipping the lamb teats in their mouths. Rejected with a splutter. Then we try coating a finger with formula and putting a milky finger in a sad mouth. Now we have interest. Letting the finger sucking get going, we then slip in the red teat and we are in business. Each little kid with trembling body starts to suck on the wine bottles.

We have only mixed a cupful for each as a start, but we will be back for our little sad babies, and we spend a few minutes stroking them. In less than a day, three feeds in, they will be frisking; in less than a week, they will be gobbling a full bottle and leaping about. We still have time, however, to creep in as they doze and stroke their fluffy ears and tell them all is well in the world.

Goat babies

The goat babies have fallen into a heap, exhausted by all that skipping and prancing. Two sets of twins curled up around each other, it's hard to tell which body belongs to which head and whose little hoofs are poking out into the grass.

Georgie and Morwenna crouch down beside them to stroke them and a bleary eye is opened and swims shut again. Morwenna scoops up a bundle of floppiness, its legs dangling, and nuzzles her face into the fur. Georgie does the same, each breathing deep the smell of newborn creatures who are too sleepy to wriggle.

One kid struggles awake and lets out a mewing bleat and his mother hurries over with an answering sound and checks out her baby, held tightly in Morwenna's arms. She puts it down and it runs to suckle, butting its little head into the udder and flicking a small tail as it pulls down a suck of milk. We stand and watch.

Pink plush

To convert them, first we must empty the farm buildings. We hardly know where to start. The kids have been using the upstairs of the small barn as a playroom for the last few years and twice we have had short-term illegal dwellers, friends who were between homes. As a result, the upstairs is strewn with a mix of toys and furniture. The furniture is chiefly a three-piece suite in pale pink plush, an Art Deco oddity with the backs in a sculptured shell shape and the arms as solid as a pale pink pig. The springs sag, the plush is threadbare, and we have no further use for them. They have to go. In fact, their departure could mark a serious beginning to the Project.

A family activity develops, getting a bonfire prepared. More challenging will be moving this hefty set of sofa and chairs out of the top level of the barn. Michael and the girls position themselves up there to consider the next step and then go looking for big ropes. We manoeuvre the first of the chairs towards the door, which leads out over a slate sill and into thin air. Michael has tied ropes to the stumpy feet and lashed the whole body with a tangle of more ropes and is taking the full weight; the girls each do steadying with their smaller ropes.

I am below, wary for my life. An enormous pink body emerges from above and slowly inches down, until I can also support it and help it to the ground. Triumphant and seriously out of breath and energy, we all take turns to sit on it, and then push it into the back of the van and drive it the short distance to the bonfire in the vegetable garden.

Morwenna is given the lighting task and deftly strikes match after match to flame up the buried newspapers. In a few moments, the fire rages up and we stand back to consider the fate of the pink plush chair. It seems to be ignoring the flames underneath it and we look on a little concerned that our labours have been pointless.

A small smoke plume emerges from the seat, followed by a more convincing smoke signal, and suddenly the whole seat bursts into flame, vigorous, orange and bright.

Childcare

I am adrift, trying to be a mother and a wife and a farmer and now a student. I am taking a degree course in English Literature, for my soul. My childcare arrangements are all over the place. I can't always get to pick up Georgie from primary school. She is walking home with Phee's mother on some days, and is at other times driven home by Martin, who is seventeen; he cooks her supper and is later sent a Mother's Day card by her for his pains. Phee and the dream team scamper in to cook multicoloured drop scones any day they feel like. Morwenna gets collected by me on some days, and on Fridays I gather up two Wilkins girls as well, for a family swim. There we meet Georgie and the rest of the Wilkins and we swim and play and rejoice at the ending of the week.

I have forgotten or been late more than once. I am gritting my teeth and will get through this; student days are brief, are glorious, make me dizzy with happiness. At least the goats and the supportive husband are coping just fine. Georgie looks squarely at me and tells me I have never ever cooked a cake like a proper mother.

Goat hooves

The rain won't stop raining; on and on it goes. Walking across the lawn is impossible; feet squidge in deep and make a sucking sound as they pull up. The goats ought to be kept in the sheds, I think, which will make them cross and use up a lot of hay, but I hate to think what it's doing to the land, or to their hooves, for that matter.

When Michael comes home, I suggest we bring the animals in early and look at their toenails. It's everyone's worst job. We arm ourselves with secateurs and Michael holds victim number one, always starting with the boss, to show the others that this is okay; he takes her firmly by the collar and wedges himself up against her and the wall.

I bend the first skinny white leg over, so the hoof points to me. It's muddy and it's also curving over itself. I clip a two-inch piece off each side, thick, white and not unlike a cut through a cooked chestnut, and the hoof then shows clean with a pinkish inner part and a now manicured tough, white hardness on each side. Both front hooves done and Michael budges along, so that I can crouch at the hindquarters to reach the back hooves.

Things are not so good at this end, there is a smelly ooze from the side of the hoof, foot rot, from damp and soggy soil. I cut back as much as I can and then spray a purple chemical treatment on all four trimmed hooves.

My back is aching and the stink of the rotten, seeping hoof goo is hanging in the air, but there is no point in stopping after only one goat has been sorted. Time to capture the next one.

Goat kids at college

We have orphan goat kids. And I have to go to college, it's a big day for lectures and I simply can't miss it. Michael takes the girls and I drive the van down to the yard and lob a bale of hay into the back, split it open, and make a hay nest for two very small, soft goat babies. I pick them up from the shed and post them into the van and drive up to the house to collect my college files and a handbag, plus milk powder, baby bottles and a made-up bottle of warm milk.

In the carpark at college, I squeeze between the van seats and wriggle into the hay and feed each of the kids, who are super-hungry and gulp the milk down in greedy bursts. I get to my first lecture, and notice a few strands of hay about my person as I sit taking notes.

By midday I am free again and I am met with piteous bleating in the back of the van. I measure out milk powder into the bottle and nip out to the Students Common Room to fill it with water. I ask the staff at the bar if they could microwave the bottle to warm it and one of my friends comes by, puzzled by what she sees. I explain I have orphaned kids in the back of the van, and I dash off to feed them.

The kids are delighted with their meal and I return to a seminar room for the afternoon.

There is some sort of fuss outside and I am summoned out. The NSPCC have been called to investigate why babies needing bottle feeding have been abandoned in a van. The bar staff have been diligent, and I have a lot of explaining to do and some forms to fill in. I am late for school pick-up.

I'm off, fast.

Leaves

Autumn has coalesced into winter overnight and a sharp frost has felled all the leaves in a final dramatic swoop. The dry, rustling high piles of leaves suitable for kicking and throwing have been turned into stiff castles with a glistening of frost on them. Ash, apple, hornbeam, sycamore, oak; all these leaves are down, their autumn glory dumped in browning chaos.

After school, the girls and I each take a wheelbarrow and race around the garden gathering sodden heaps and hauling them off to the yard for the animals to browse. Actually, it's all a disappointment. They make a gesture of nibbling here and there, but dead and cold leaves are nobody's treat. They will make great compost, though.

Lawns cleared, we consider how valiant it would be to tackle the vegetable garden too, but it's cold and the sun has slumped away; we have pink cheeks, stiff faces, wet noses, and there is homework to do.

Photographs everywhere

It's that time when small girls at primary school are combed and sent out with a clean blouse, and a few weeks later present us with a heavy envelope in which there are several sizes of disappointing embossed photographs. This time the results are astonishing. Both Morwenna and Georgie look adorable, looking straight into the camera, smiling and with Morwenna's arm around her little sister. I melt. We can have endless copies of these, gifts to grand-parents, treasures for us.

Phee looks at them approvingly and tells me that her father as headmaster gets a complementary photo sitting each year. 'Does he want such a thing?' I ask innocently and I start to giggle. 'Does he send one each year to his mother to line up on the piano? His beard might widen or lengthen from year to year!'

Louise's boys come home with photos of crumpled shirts and egg stains on the chin. It is a risky business.

I have this week been chasing goats and ducklings around the cottage lawn for photographs to grace our website. They at no point wanted to stop and charm the professional photographer.

I decide to do better myself and have an afternoon in autumn sunshine taking evocative pictures of cabbages, apples hanging from the tree, nettles, the goats in the field, and the river. I crouch low to click away at the chickens and am lucky with a fabulous goat portrait of Yamaha. Girls and goats clearly make better subjects.

Managing secrets

Why would I be creeping around looking furtive on a bone-chilling cold November morning? I am really searching out places in which to be secret, so that I can spread out bits of wood and paint and scissors and glue and start making Christmas presents.

The restrictions are considerable. I have to deceive both husband and nosey children. I need to be reasonably warm, free from cobwebs and spiders, free from the insistent and warm breath of goats, and have a flat surface to work on. This so far is ruling out the animal sheds and probably the top barn, which disqualifies itself with cold and spiders and dust. The milking shed is far from secret and too cold. The workshop is impossible, there hasn't been an inch of space in it these last eight years.

It will have to be the kitchen table of the larger cottage, which means I have to get cracking, because we have two bookings, one next week and one the week before Christmas. This will speed me on. I can use this week to sort hardboard and cut bits of wood in the workshop, find what paint I can use and what I need to buy, and also test glue for old age. I clear a little space behind the black oilcloth coat in the washhouse cupboard to store my essential equipment.

Twelve placemats to be made for Michael from shiny paper tote bags with details from 'The Lady and the Unicorn' tapestry. One farmyard for Georgie with cow barn, goat shed and painted pond. One set of furniture for a dolls' house for Morwenna. Three weeks of mysterious disappearing with assorted excuses: apple turning in the big barn, goat grooming, small repairs to curtains in the cottages. It's a race against time and suspicion, but I am winning.

Chapter 8

Two little bugs go out to cook

The children are voting with their feet. They have definitely had enough of winter and long for their carefree garden days. They decide to go outside and cook breakfast. They have actually had their breakfast, but these things needn't be limited. They bustle about planning what will be needed and assemble in a box some eggs, butter, and two slices of bacon. They then negotiate for a fire and we tell them they can have the Camping Gaz stove, but need to be very careful. They find the ancient frying pan from the camping box, and clearly know all there is to know, because I see matches and a wooden spatula. They put on boots and coats and hats and sail away to the best cooking place under the yew tree and set about it.

Michael and I watch from the window. They have been fighting like lionesses for the past week and suddenly this has brought them together. Morwenna is in charge, but being so kind and sharing out the jobs. She is holding the pan steady on the lit gas and Georgie is allowed to place the rashers in the melting butter and then break eggs into the pan. It is harmonious and so very lovely.

All they needed was to be outside. It feels like spring is here.

April morning with goats

Morwenna leads the goats out of the yard and up into the triangle top orchard. We are a funny little procession. This leader is nine years old and confidently in charge of animal routine. She has Yamaha by the collar and together they are considering small treats for a goat to nibble along the way, snips of young bramble, sorrel in the bank, a leaf or two of primrose. Behind these two are Michael with four junior goats. Straggling behind and chatting to herself follows our dizzy six-year-old.

I am bringing up the rear with Juniper and her small kids; she needs leading, as she would prefer to be in the shed with her little ones and they are frisking about in a new and astonishing world. They plant their peg legs squarely on the ground to snort with surprise at what they see and seek reassurance with a tinny bleat, which is answered by a snicker of maternal affection.

Morwenna opens the little gate into the top orchard and firmly directs Yamaha towards a tangle of brambles that we want stripped, so we can cut the remaining briars. We stand in a pastoral sort of way; the goats are not tethered and need to be kept focused on the bramble zone. Georgie and Morwenna flutter about finding catkins and pussy willow and a tiny hidden violet. It's a perfect April moment.

Judy and the den

Zafar, Edward and Morwenna are building a den. Morwenna is assuming a worryingly subservient female role and I try to persuade her that this is her patch and she is equal, but my feminist stance is going nowhere. Edward is the brains and Zafar the brawn, while Morwenna is the moll who fetches and carries important stuff for a camp.

It's top secret, but they need some adult input and Judy has been enlisted as a secret helper and isn't allowed to say ANYTHING. The map and plan of the camp, the farm layout and key markers, has been left on the kitchen table in case it gets muddy and the camp builders have gone roaring off to find a store of logs and plank and rope.

As I pick beans in the garden, I can hear rustling and whispers in the laurel hedge, and when I look round there is a wooden ladder propped up on the stone wall below the hedge. I might be tempted to think there is a camp close by. A small figure hurtles past with a frying pan and a milking stool. Georgie comes along whining that the big ones won't let her play, so I say to her very loudly that it's a pity, as she could be a very useful member of the camp who could take ice creams into the hide-out. We go back to the house.

The camp leaders soon arrive and whisper to Judy, who then takes Georgie by the hand and they carry four ice creams along the secret path through the secret orchard and into the Top Secret Camp.

Goats need an electric fence

I'm in a fury of guilt and remorse. The vet has been called and has cast a broken leg in a long, white sausage of white plaster, covered with a sticky bandage, and I am blaming our no-longer-fit-for-purpose method of tethering goats. When we had only a few goats, it was possible to have each one with a collar and a rope running between two metal stakes, and if there were babies, they could dance and jump about as they pleased and their mothers could shout to them. But we have more, many more and although the babies still run free, the teenagers are tethered alongside the adults and the Goat Who Must Be Obeyed, Yamaha.

Yesterday there must have been an argument between teenage goats and troublesome bullocks, and now we have a sad goat, Conker, with a broken leg. I suppose an option might have been to despatch her for future roast dinners, but we had been planning her career as a mother and grandmother, and in a flash, I had called the vet and been told to bring her into the surgery.

So, I have taken the bold step of buying the kit for an electric fence. There is a battery on a little stand and it connects to wires that are threaded through insulated fencing posts. Michael seems very pleased and we set up a corral in about half an acre and let the goats in. They are initially rather suspicious, but Yamaha is not one to mess about and she plants her feet firmly in a patch of lush grass and starts to eat. The others look about and do the same.

We watch them for a while and feel like happy parents, but are also curious to linger on to see what happens when the electric fence is touched. Did we switch it on? They don't seem interested in testing boundaries, so we walk back to have a cup of tea. Shortly, we hear a little shocked bleat. And another. We figure the boundaries are indeed in dispute and look over the ha-ha to witness a goat discussion group.

Whatever the group decided, it was deeply intelligent. They will never touch the fence again.

Managing waste

By the sink is a full, shiny green bowl, pretending to be an exotic salad, but it is actually stuff waiting to be compost, or hen food or pig food or all three. It contains a heady mix of peelings, crusts, stalks and slops. I like to think the carrier of this bowl is specially honoured in the yard, the bringer of treats and delights. Who knows what the geese or ducks are calling when they see the green bowl? I dump the contents in the pig trough, or on the concrete path, or on the low mountain of manure out in the field.

The manure heap will rot down over a year, taking with it any fibrous rubbish, whole stems of Brussels sprouts, bean stalks, floppy rhubarb leaves. The vegetable garden compost box will have reduced the weeds and lawn clippings to a slimy goo and the day comes when trenches are dug and the whole manure heap gets churned up and spread over the garden. It's part of the annual cycle and it's hard graft. Gloves must be worn to minimise the blisters on hands, as the heft of the shovel is cruel.

I like that our domestic life offers itself up into the trenches. Here goes a rubber band, like a long dead worm, or one of Morwenna's old Baby William bean babies, now with a soggy cloth body and smeared little face. A wine cork. One of the goat collars. My missing glove, a silver teaspoon. Is it too much to hope that all the lost things creep here? I would like to find the lost credit card that made Michael so cross.

Glut

The girls have collected eleven duck eggs this morning and in the fridge I have another six and a good many hens' eggs. I need to distribute some, but first I think we should use as many as we can. I have some strategies:

1. *Twenty egg cake.* The recipe is simple. Weigh the eggs and then measure out the same weight in butter or soft margarine, sugar and self-raising flour. The girls can probably do this with their eyes shut. Mix, mix, mix the sugar and soft butter (to a cream, my grandma used to say) and then beat in the eggs. Only stir the flour, as there is raising agent in it that mustn't be beaten, and then add water or milk (to dropping consistency, said my grandma).

   Flavour with vanilla, or chopped apples and cinnamon and raisins, or lemon zest and lemon juice, or splashy wet coffee granules. Then grease and flour the two meat roasting tins and spread out the cake mixture and bake in 180°C for 30 minutes or until no longer soggy to the touch.

Lick the spoon if it was hen's eggs, don't if they were duck. 'Why?' asks Georgie. I remind her of the unhygienic state of the duck shed.

2. *Fewer eggs for home-made pasta*: 1 egg to 25g of plain flour, some salt and lots of kneading and then fun and games with the rolling and cutting machine with its hand-turned winder. Drape bandages of damp pasta over the clothes drying rack and the backs of kitchen chairs. Slice into ribbons. When it's not quite dry, drop the pasta into a huge vat of boiling salted water and cook for 2 minutes. Or save it floured and soft in a poly bag in the fridge.
3. Separate eggs and make mayonnaise or lemon curd and meringues.
4. Find mums and friends who need eggs. Sort out a supply of egg boxes.

Wait. I have a contract! I am to take as many surplus eggs as I have to the wholefood restaurant. Sorted.

Eggs for sale

One of the best things about being organic and self-sufficient is bartering and selling things. It is with great pleasure that I take a tray of eggs to the wholefood café, the Rainbow Café. It is run as a co-operative. Hippy characters take turns to cook, wait at table or serve food from the counter, or run the crèche in the room alongside. What isn't so good is my battle with the bad-tempered traffic warden, who seems to wait for me and pounce. We have the same conversation most weeks.

'Madam,' he leans into the driver's window sneeringly, 'Madam, you may not park here.' I reply wearily and patiently that I know it is acceptable to make a delivery.

'But you are not a trade vehicle, madam, or making a delivery.'

'Yes, I am,' I protest. 'I am carrying a tray of eggs to the Rainbow Café.' And I get out, attached to a child, and bear aloft a battered cardboard tray of eggs. He looks at me sourly, again.

On this particular day, my battle is of a different kind. I have only a half tray of eggs, mostly duck eggs in a gleaming blueish white. When I hand them over to the cook in charge for the day, I apologise that there are so few hen eggs and by way of explanation I say that two of the hens are broody.

I am met with a stare. I say cheerily that two of the best layers, the dark Marrans, have each got a clutch of fifteen or so eggs under them, looking smug. I get the blank stare again and am nonplussed. The cook says very slowly and carefully with an appalled look, 'You mean the eggs are FERTILE?'

'Well, yes, of course,' I reply.

'Oh, my goodness, this is a vegetarian restaurant,' she says. 'I will have to tell the committee that your hen eggs are fertile.'

'And the duck eggs,' I throw in.

'We are having a meeting on Tuesday and we will call you to say if we think we can use FERTILE eggs.'

I roll my eyes as I leave, with my rejected fertile tray. They want virgin eggs? Let them have battery!

As I turn the corner towards the car, I see the warden. Am I no longer a valid delivery? I feel the need to hide a while at the pet shop window, examining little hamster toys until he passes.

Boats

Michael has bought a surprise boat. We are now to be a sailing family. I thought we had knocked that one out when we sold the Fireball dinghy, a supercharged monster that flew like the wind and had me strung out on the end of a trapeze wire. It had ended because I was too pregnant to do up the trapeze harness, and I objected to being ducked into the water by the smallest twitch on the helm.

Now we have a boat again and she is parked in the haybarn under a canvas covering and there is a plan and a threat. We will go to the Isles of Scilly, but not until Georgie can swim a width in the municipal pool. It has tempting possibilities, camping on an island without cars, endless sandy beaches, lobster to eat – a long, hot family summer. But what about the animals, the weeding, the harvest? Nothing could be less compatible with a sailing summer than living on a small but labour-intensive farm. Michael says it will be fine.

Georgie puffs and pants her way across the pool and gets her 25-metre badge. I sew it onto her costume. Who is going to look after the farm? Michael goes to work. Who is going to look after the farm? Michael books tickets on the ferry for four passengers and one boat to sit on deck. The situation shifts and tightens. Who? When? And then with a mysterious click of chance it is all solved.

My Dragon Aunt is thrilled to offer the services of herself, her collection of grown adults and their small children; it will be their family holiday.

She arrives with a few days in hand. Her team follows at intervals, spilling out onto the drive in full holiday garb: Hawaiian shirts, sunglasses, flip-flops, it is Cornwall after all and they ignore the drizzle.

We hand out the colour-coded list of chores, red for milking, blue for watering, yellow for food for which animal and in which shed. 'It's foolproof,' we say and we drive out the next morning singing at the top of our voices. The boat trailer loses a wheel as soon as we join the main road, but we stop and fix it. At Penzance, we watch as the little lugger is winched on board; we are drunk on the joy of escape.

Tooth fairy

Georgie and I have spent some fanciful moments perched on the edge of the flowerbed reading a tiny message from the tooth fairy. It is written on a page the size of a postage stamp and has clearly given the writer some difficulty, but it says some nonsense about life in the flowerbed and signs itself 'Wisteria'. Georgie is enchanted. She has recently received two shining pound coins from the generous Wisteria and is now enjoying a flowerbed dialogue.

Louise arrives and Georgie tells her that there is a letter from the tooth fairy and Louise breaks out into a guffaw and says, 'You don't still believe in that nonsense, do you?' Georgie looks at me to reassure her and asks, 'It's true? There really is a tooth fairy, isn't there?'

This is a bad moment. Do I tell a lie? I wear an expression of extreme sorrow and tell her that no, there isn't really a tooth fairy. She takes a moment before collapsing in hot, anguished tears and races indoors and away. Presently she reappears and gasps, 'Father Christmas? Jesus?'.

Massage parlour

It's peak summer holiday and we have Richard and Judy and Ed staying and it is pouring with rain and has been for three days now. The builders who are working on the milking shed are busy with the roof, poor things, and we exchange cheerful words as I go past with some eggs.

We have just about run out of things to do inside, everyone is feeling cooped up and in need of an adventure.

On the lawn the big family camping tent is up and empty, and I dart out to it with an armful of groundsheets, tarpaulins, and a sheepskin rug. I need more things and I dart back again through the rain. The children look up briefly from a game on the kitchen floor, but I am not attracting attention, yet. I have sarongs, string, an electrical extension lead, pegs, a fan heater, some blankets, cushions, an umbrella and a bag of stuff.

Finally, I make an announcement. There is a massage parlour in the tent in the garden and guests may book their slot. Mostly there is a baffled reaction to this, but Morwenna jumps up, eager to have a go.

We hurry through the rain to the tent and I escort her to the changing-room section and tell her to drape one of the sarongs around her and call me when she is ready. I then lead her into the massage room where the fan heater is making a tropical breeze and blowing the sarong curtains about and I give her a back massage on the raised plinth of cushions and sheepskin rug. I escort her back in a queenly manner under the umbrella, ready for the next customer.

These are not hour-long massages and I have a fairly quick turnover and a steady stream of customers. The builders are frankly curious to know what is going on, but we remain aloof and mysterious. It is finally Richard's turn. He is reluctant, but has received much persuasion from the satisfied customers indoors and I am persistent. He is terrified of revealing flesh or, worse, underpants.

I lead him out, drag him out, holding his hand, and he shouts loud protests for the attention of the builders. When I show him the changing room and the sarongs, he emits a volley of noisy but unconvincing complaints, 'We can't do THAT, No, No, I won't. You can't make me. No, No!'

One of the builders calls out, 'What's going on? Do you need help in there?' Now Richard and I collaborate in leading the builders on with weird, misleading cries and shrieks. Then we hear a scuffle and the sound of falling roof tiles and a loud yelp and we agree it's time to slip away; enough entertainment for one rainy day.

•

This is something I can't abide, but which Michael and the children love, so I will join them only when there is a crisis.

In a smelly, polystyrene incubator, eggs are hatching. They could be duck, goose or hen, or even occasionally guinea fowl, but every day they need to have a spray of water to mimic the absent mother's wet feathers. Every day they are turned and the temperature checked on the thermostat. Then there is the bothersome/joyous moment when the spray is met with an indignant squeak from inside the shell. After that it is only a matter of hours or days until cracks appear in the shell and a tiny little sharp tooth on the beak pokes through the shell, and a bleary wet head appears.

It can take a day for a chick to shake itself free of its egg, troubled by its outsize feet and the newness of life. There is no mother to fluff up its feathers as this is incubator work and so the chick looks dank and greasy, until Michael and the girls set up a lamp and a box and sawdust or hay and then the chicks start to dry out and transform into proper lovable creatures. I can join in from now.

There are some egg failures, with accompanying stench. The slushy mess that the babies in the box make with their given meals of boiled egg mixed with chick crumb is also smelly and in no time it is caked into their feathers. They really need a mother. If we can slip newly hatched creatures, by night, under the feathers of a mother-to-be whose hatching has failed, that is surrogacy at its best. The bond is instantaneous and a broody hen whose sitting has long overlasted the viability of her egg is thrilled with her new duckling children; she doesn't yet know the teenage problems ahead on the edge of the pond.

Those hatchlings, without the benefit of an adoptive mother, need to make do with Michael and the children and lamp and box. They clamour with a welcoming noise and are taken out and caressed and eventually allowed out to romp on the grass in a cage, to keep them safe from magpies and crows.

By the time they are old enough to become part of the farmyard, we are anxious for their wellbeing; will they survive out in the wild without their gourmet meals and bedtime stories? We make the move at night, placing the young ones in with their older and unfamiliar relations, and hope that in the morning they will be ready for a new life.

Georgie and Bethany outface two large bullocks, Sweet William and Sweet Pea

Georgie and Bethany have taken themselves out, in the way important seven-year-olds do, to strut and point out things along the riverbank. I watch them set off with a basket over Georgie's arm; who knows what it contains.

They return pink and flustered and the basket is missing. Together in broken sentences, where each overrides the other, they have a tale.

Georgie starts, 'We were on the bank, talking to river fairies and getting daisies—' Bethany cuts in, 'They saw us and started running . . .', and Georgie takes over to clarify, 'Not fairies, but Sweet Pea and Sweet William, galloping. We thought they were going to knock us down.'

Bethany sticks her chin up, 'We weren't frightened,' and their words tumble over one another.

'We knew what to do . . . We stood still and put out our hands . . . We shouted, "Stop! Stop!" . . . We made ourselves look big . . . "Stop!" . . . And they did, they just stopped in front of us and started huffing, so we patted them, one each.'

'I did Sweet Pea,' adds Georgie.

Testing the water

Part of the planning conditions for our holiday cottages is that the spring water has to be tested and today the man is scheduled to do it. I have got Owl Cottage kitchen scrupulously clean and rubbed antibacterial wipes all around the tap. He arrives and brings out a little kit, which looks like a portable alchemy set.

He swabs the tap and then brings out a bottle of meths, and with it he sets light to the tap to kill off germs and then pours a little water into a bottle. As he does this, he bends very close to the tap to check where the hole and bottle meet. He seals the bottle; job done.

I am standing by, looking puzzled, and he explains that the chances of bacterial contagion are very high, germs could be lurking in the tap, in the water (and in his bottle and swab too, I think, to say nothing of his hairy nostrils, which have been breathing into the tap), and the test result must have only a miniscule level of contagion to be passed.

I don't like the idea of contaminated water. It sounds worryingly haphazard and I think of the spring supply in the squelching, soggy cow field dripping into our collection tank and feel that there is not a chance that the water will pass its test. We are perfectly fit and have adjusted our insides to the local bugs, but the thought that holiday guests will be poisoned is too horrible. I wait for the result.

Wet creatures

It's raining stair rods. I rush out to see to the animals, in a big black coat, hat pulled on, head down, scurrying along fast.

The goats announce that they will be staying in, thanks, and can we have more hay in the manger, please? The hens are too stupid and rush outside to gobble scattered grain, even though we have thrown plenty onto the hay-strewn floor, where they could scratch about for it keeping dry and warm. The ducks and geese race out, unperturbed.

We bring a bale of hay and throw tranches up into the manger and we hurry back to the house to start the normal day of most households: breakfast, bags ready for work, lunchboxes, teeth cleaned, suitable clothing.

It rains all day. The roads are flooded and cause delays, and such unstoppable rain makes me worry about our leaky house and I wonder if we have drips or worse.

There is quite a flow of water running off the fields as I drive up to the house. Our drive is awash and the diverting ridge we had made has been washed away and water is flowing into our drive instead of down the hill. First things first. Change into farm clothes, wellies on, waterproof trousers and jacket, and sprint into the yard to get a hoe to skim a thin channel between our drive and the concrete road to send all the run-off water down the hill.

As I am waterproof and the rain is like Noah's flood, I decide to put away the animals early. What a pitiful group of goats greet me. Their stable door has blown shut and they have been excluded for some time from their warm, dry, hay-filled shed. They are outside, drooping and complaining. I point out to them that there are three other sheds in which they could take shelter and there was no need to be soaked to the skin. They each look three times skinnier than when their fur is dry.

We go in and I take handfuls of hay to rub them down. They eat the hay furiously, as if to make up time, and then gobble their oats. The hens are sodden too in their shed, feathers soaked and sticking flat to thin bodies.

I walk out to the meadow and call the ducks and geese. One group is paddling in the reeds, a thin line of ducks sailing down the river like royalty, others grazing or glugging for snails, but at my call they look up and start an orderly line towards the yard, calling back to me with quacks and honks. Each and every one is pristine, gleaming whiter than white.

Sweet William to market

Today I was in a fluff of nerves. It was Jack Kitto's fault. 'You going to take those bullocks to market?'

I bowed to his superior wisdom, it is probably time. We will keep Sweet Pea for the abbatoir and send Sweet William to market. But what a performance. At 7 a.m., Michael and I left the girls having breakfast and a promised piano practice and we backed the van close to the yard entrance to load up Sweet William. Now the moment of hideous betrayal was to begin. Sweet William trotted in pursuit of the oats right up to the yard gate and then had a moment of doubt at the unfamiliarity of this new business.

The gate was open, the van back door was open, the wooden gangway up into the van was down and I hurried up into the van with as much oaty encouragement as possible. Sweet William let the greed of the moment seal his fate. He was up and in and Michael shut the door with us both in the van. I left Michael to get the girls off to school, and quickly threw on Farm Jacket No. 2 in deference to the gaitered and hatted farmers I would find at market, and drove off up the hill.

The van lurched and wobbled at every corner as Sweet William shifted and stumbled in the back of the van, all the while shouting, mooing, slathering nervous dribble and pouring sloppy excrement onto the van floor. My heart was pounding with the anxiety of it all.

At the town market area, a guy in a white coat looked helpful and directed me to some makeshift animal pens. There a cluster of older men, white-coated and knowledgeable, told me exactly what to do and how to do it. Sweet William was glad to be let out, less so to be manoeuvred into place by his tail. I hurried to register our presence.

The auctioneer was most impressive in his tweeds. I told him I was inexperienced, had never brought an animal to market, and needed instruction. 'You want to lead the critter around the ring?' he drawled and I quickly replied, 'No, very definitely not, please let one of the yardsmen do it.'

To my astonishment, there was talk about our likely receipt of a subsidy. For goodness sake, growing Sweet William on from the trembling, paltry unwanted £5 calf from this very market two years ago, and we were to receive £50 from the government for doing so? Astonishing.

I sneaked into the covered arena for the auction and watched the process. Many of the local farmers I knew were chatting and lounging in the ringside seats, half observing the beasts as they were brought in, and half enjoying the rub of fellowship. Sometimes the yardsmen would poke a few animals in together; sometimes an owner would lead a creature around the ring, and the auctioneer's song of 'What am I bid?' in its high-pitched tones, almost indecipherable except to the practised ear, would accompany the bovine parade.

The farmers seemed to be impassive, but for the most discreet of nods or winks, well-known to the auctioneer, who kept the performance rolling along. Then it was Sweet William's turn.

The auctioneer turned towards my hiding place in the arena and called out, 'Mrs Hodin, it's your bullock now, will you come and lead it round the ring?' The brute.

All eyes turned to me and I had no choice but to slink out and take my shuffling friend by the baler twine and lead him around. I got cat calls, whistles and ribald remarks; Sweet William and I made several circuits with me stroking his neck for moral support. He looked good, rippling fat and with a sheen on his coat.

We got a very good price.

Chapter 9

New pond

We need a pond in the field so that the new ducklings can learn to swim in nearby safety, instead of their perilous lessons in the fast current of the river. There is a slight dip where a spring occasionally drains into the field and that seems a good place to start. If we dig deeper into the dip and pile up what we dig onto the sides and widen it, well, that has the makings of a pond. So we reason.

The digging is a full team effort, the girls wielding shovels and chucking stones and earth onto the sides, Michael and I more prolific with effort, but arguing over how deep and where the rim of the pond is to be. I'm accused of being short-termist and a bodger, Michael liking things done properly. Morwenna tells us to stop arguing, and we dig and throw, dig and throw.

Georgie takes on the job of stamping on the edges to firm the soil, singing a small song as she goes. The hens are interested, believing that we are worm hunting for them, so Georgie's job expands into shooing them away from spade work. Goats have come and gone.

We tire of our work and decide that a great depth is not really necessary. We stamp on the bottom to get a solid base, and drag a long hose through the yard gate and out to the field to test the shallow pit. The hose doesn't reach the pond. A relay of two watering cans and carriers pour and pour and pour, and we watch the water rising unbelievably slowly. We have exhausted most of the labour force, so resolve to finish filling it later. We let out the ducklings.

When we go out to put the animals away, I imagine a pretty scene with ducklings enjoying this close amenity. Instead, it looks like an Easter chocolate factory, with shiny brown ducklings arranged by a pool of thick, gooey mud.

Running away

I'm running away, maybe for the hundredth time in our marriage. Michael is impossible, growling and snarling and won't discuss why. I can't bear a second more. I used to jump on a train and go to London, stay with Clarissa, but I have two kids that have to get to school, and who certainly don't need to be run away from.

Over the years my running away has got shorter, to now a pointless activity of revving the engine and departing in a dusty cloud, up to Liskeard, finding nowhere to park, coming back past Barking Dog Farm to take a bit longer, and then fuming in the forest lay-by at the bottom of the drive. 'Back then?' goads Michael.

He used to diffuse the situation with nervous laughter, once standing in front of the car into which I had locked myself as I revved and revved and revved, until suddenly smoke started pouring out from beneath the bonnet and our eyes locked in horror. I jumped out and he laughed and then I laughed.

Today I tell Morwenna we have to go away for a few days. She asks why.

'Because Pa is too cross.'

She nods. She has been running between us and has seen the tension growing.

'Where will we go?'

'To Liz,' I say, 'she is close by, you can still go to school and Pa won't think we will be with her. Pack night things and school things; I will do Georgie's.'

I ring Liz and check we can stay for a few days; she asks no questions.

By chance, Judy rings for a chat and I spill out my storm to her . . . 'horrible, unreasonable, beastly Michael' . . . and she laughs. I tell her we are running away to Liz for a couple of days and she laughs again. 'Rosanne, not having you around for a couple of days will feel like a holiday for Michael. You need to disappear for weeks before the pain will hit.'

Aargh. I can't stay with Liz for weeks. But she's right. And I do not want to give Michael a delicious Rosanne-free holiday.

I take the girls with me on a long walk in the woods; when we get back, we are all feeling better. We unpack.

Goats for Christmas

We are going away for Christmas and can hardly expect anyone to come over to milk goats. We once drove two goats down miles of bumpy farm track to be cared for over a long weekend by a strange lady goatkeeper, with ragged clothes and a beard. We came to collect them mid-afternoon and by the time we had loaded them into the van, the small winter sunlight had faded and the world had refrozen and we couldn't get away over the ice. We had to unload the goats and stay over in her house in a room colder than the grave and we ate goat chops. She was so kind, but it was ghastly.

It is a Cotswolds Christmas this year and I have carefully negotiated with my parents that it is okay to bring up three goats that will need milking. 'Perfectly fine,' my mother has said in a wavering voice. I don't think she can quite visualise the scene.

I pack up the car with Christmas presents and suitcases and children. Michael packs up the van with bales of hay and food bowls and sacks of oats and various lengths of string. It all seems reasonable. I leave. This is before the days of mobile phones, so we have arrived and played and had macaroni cheese and bath time and still Michael has not appeared. He finally turns up with an injury he cannot fully explain, but his thumb is twice its normal size with a throbbing purple nail – an accident with a hammer. I make him take a glass of whisky and offer to pierce the nail, but am refused. We start to deal with the goats.

My father and I have moved stuff from the little garage and so we fling the hay bales into a corner and I untie one and break up the hay into a cosy bed on the floor. The icy air makes clouds of our breath and the valley is silent. The goats tumble out, having settled down to sleep for most of the journey. They allow us to milk them and feed them and shut the door. But all of this is too strange for them and as we adults are sitting and eating dinner, the call of our baffled goats bleating can be heard up and along the Daglingworth valley.

A sort of Christmas carol.

Transformation of the yard

It's unbelievable; we have complete yard transformation. We have covered up the grime, the pawprints, the mud, the gravel, the grit, the slow dribble of water that thinks it is a stream running diagonally across the yard. It's all covered up. We now have a space that will soon be grassy splendour.

We have captured the stream dribble into a slate conduit, nicely covered over in case it one day decides to flow faster than its current sluggish rate, and we have brought in lorryloads of topsoil. The lorries have dumped the earth in miniature pyramids, and we have leapt on them and dragged back the soil with rakes and spread it evenly over the yard. It is neat and clean and uniformly brown. It's amazing. It has taken hours. My back is burning like fire.

Next we have walked, oh so carefully, up and down with footsteps that feel like fairy feet and have scattered grass seed thickly over the earth and raked it in. The animals are forbidden, banished into the fields and only allowed to wander the two-metre-wide concrete pathway that surrounds the newly seeded courtyard, and to look in nostalgically.

Offset by this smooth brown place, the two cottages look elegant and trim, the big barn strong, secure and full of future promise, but the corrugated-metal-clad food barn looks an utter disgrace. It's always one step at a time. Today we have brown courtyard. Tomorrow we might have grass.

A mating party

We must take Rosetta up the hill to woo Jack's bull. I will never have a repeat visit from Mr Insemination from the bull farm; years ago his only trip to offer Rosetta the gift of pedigree semen was a brutal rape with a long-gloved hand, and he will never set foot on our land again.

His blast on the car horn woke the baby. He unloaded a blue bucket containing straws of frozen semen and then put on a rubber glove that carried on up to his armpit and started a military march down the drive with wailing Morwenna and her very cross mother at his side.

I had to hold Rosetta's collar while he picked up her tail and plunged his arm and the straw further than I could believe possible. She bellowed, Morwenna howled, and he stomped off back to his car, where he peeled off the glove and dropped it on the ground and demanded his fee. I wanted to kill him, so great was my fury. I had been the one to invite him and his service.

But Rosetta did throw a beautiful calf.

Today there will be no artificial insemination. Jack is happy for his bull to play his part. We are to walk Rosetta up the hill.

Michael is in charge and lashes on a leading rein and then has to restrain her; she is walking faster up the hill than he can. It feels like a carefree country morning of bees and sunshine and spring flowers. At the cattle grid, we all stop and wonder what to do next. Obviously, we are not meant to pass: there are metal cross-bars set into the ground to prevent Jack's cows from escaping. Rosetta pauses only for a moment and then reckons she can creep along the side, dragging Michael behind her.

As we enter the first of Jack's fields, we meet a suspicious sentry: a cow who has been watching our arrival. We are not part of her herd. She and others come forward to watch us. We are walking at a brisk pace on the concrete drive. The cows form a half circle ahead of us and one snorts and steps forward to look menacingly at Rosetta. But this oestrogen-rich cow of ours is not going to be put off; she has a powerful mission. She ignores their hostile looks and pushes on.

We don't know where the bull pen is, but Rosetta finds it quickly enough, and is standing beside the metal fence looking in when Jack arrives. His bull has already greeted Rosetta with an earthy mumble and she slips in and they circle each other, sniff and nuzzle until the bull steps back, arches and lunges up on Rosetta, then shudders a powerful thrust into her. We stand around and watch. Jack is pleased to see such masculinity and remarks that his bull is a proper job.

The bull and Rosetta seem blissfully unaware of the prying audience and nuzzle each other in a romantic way and mate for a second time before Jack releases her, quickly shutting the pen before the bull can follow his new sweetheart.

In the field, the cows approach again but Rosetta, her head held high, gives dainty little prancing steps and they let her pass. Respect.

Cornish hedge

When is a hedge not a hedge? To drive down past Bolitho and Barking Dog Farm is like entering a tunnel. The hedge rises high above your ears, bursting with pink campion, or primroses and bluebells in the spring, and then in high summer the grasses, rosebay willowherb, and docks stretch out trying to meet in a caress and any car hurtling past gets a pollen-filled smack on the windscreen.

The thin road twists and turns and there is room only for one. At night it's easy, any rare oncoming car will fill the tunnel with a shaft of light, and you can slow down and try to work out who has the closest passing place. In daytime, you're in the lap of the gods. Woe betide any fool who thinks that a nudge into the soft mass of campion is the solution. The hedge is a monster and always wins. It is made of lumps of granite packed together with soil and is topped with a duvet of turf from which any plant may choose to make its home.

We are struggling now to remake the hedge that slithered into a wide mess on the lawn. Our cowboy builders hadn't understood the Cornish hedge principle and had merely raised two parallel walls of stone and filled the middle with earth. We are picking through the avalanche to find stones big enough to be the crucial locking tie pieces that will hold it together.

We have made a crude wooden form to hang over our structure to show the right wedge-shape and the height, keen to prove ourselves better than the builders. It's back-breaking work, and we are squatting to choose the next piece and trundling back and forth with earth from the vegetable garden as the packing between stones. Surely there is enough stone here to remake the hedge, but we are short of the big tie pieces, and we go scouring the yard and boundaries looking for possibilities.

One dedicated day and so far we two have yielded only a long base layer of less than a metre. This is going to be one hefty project. I start to think in terms of a working weekend, lots of friends, fun, food and chat, and several jolly hours of wall building. Maybe? Just maybe?

Water issues

There is brownish stuff coming out of the tap. It's the best spring water in the world, but something has gone wrong. There could be hideous solutions ahead: to replace the water pipe from our spring; to sink a borehole on our land; or to spend a fortune digging a mains route under the railway line . . . no, no, no, that would bankrupt us. Having holiday cottages will increase our water use and it seems that we have an inadequate supply. It's baffling, as this is a busy, working dairy farm, and cows always drink bathtubs of water.

The first step in our water management is always to trudge up the hill to Robert's field, where our spring rises, and inspect our water tank. Michael carries the biggest heavy-duty screwdriver on the planet, the sort Dan Dare would use, and has a crowbar in the other hand. The female section follows with torches.

The tank is set partly underground, with a concrete top not unlike a wartime bunker, and close by is a cattle trough. The two should never be so close. The cows trample on the ground and get it soggy and somehow water seeps into the tank, spreading mud and contamination into our supply. We have long had a dialogue that the trough should be elsewhere and fenced off.

Today we shoo away the cows and Michael levers off the metal cover and we peer in. It is nearly empty. There is a slow drip of water feeding into it and the residue is looking pretty murky and the drip echoes as it splashes down. This is our drinking water.

We shine the torch around to check for cracks and leaks and stare down into the abyss feeling gloomy. The supply is obviously inadequate in every way, yet the water from the tap is delicious and sparkles. We shut the lid and sit on the bunker. Eventually we agree to come up again tonight to see how much dripping has filled the tank.

We walk down to the wood following the obvious route the piping must take; it is not long before we find enough sections of old metal piping poking out of the ground with rusty splits to know that the problem lies in old piping. Water is leaking out and mud is squeezing in. The squelching cows are not to blame and it is time to replace the whole system with new alkathene pipe.

It will be a horror story of cost, but at least we don't have to dig under the railway line.

Hair

Grooming goats is easy; their hair is short, quite brittle and responds easily to a stiff brush, and the goat receiving such a treat stands still and nuzzles appreciatively. Stroking baby goats is pure delight.

Grooming our daughters is more complex and far less satisfactory. Morwenna's hair is madly curly, like Michael's, but she wants it to be long, straight princess hair and struggles to brush it straight without the least understanding that this is going to be impossible. Today I have watched her busy with the hairbrush and her shoulder-length hair is sticking out at almost right angles like a gorse bush.

I try again to explain the folly of her ways and persuade her that if it was cut, it might behave better. She considers the situation, watching Georgie's attempt at long, straight princess hair by wearing multiple pairs of my tights on her head with the long legs and feet dangling down, in such a way that she can toss her head and feel a great weight of tresses.

I think I have assent and get out my sharp gold scissors and snip swiftly before Morwenna can change her mind. She now has a crown of heavenly curls at ear level and I stand back to admire my artistry. But I am a fool to bring her a mirror. She says nothing, simply stares and then vanishes upstairs.

She won't come down.

Gathering moss for Easter

We are on a foraging party with fancy baskets on our arms. For Easter we will make miniature Easter gardens of moss and small stones and twigs and flowers. We have found suitable tin trays of equal size and now we must get our greenery. We pace up and down the drive looking first for moss. The vertical slabs of slate that form most of the bank on the drive have coverings of velvety moss, like on an antler, but they are impossible to scrape off. We are disappointed and set off to the woods with serious intent.

We find mosses in squat, furry mounds, in spikey tree-like shapes and lush wet clumps, and we harvest all of these and put them in the baskets. Other booty is here, too. Buds of pussy willow, sprouting twigs of sappy leaves, some catkins and a few primroses, and we march back eager to start the project.

One Easter garden has a mirror edged with little stones to make a pond. Pennywort has been clustered in one corner to look like a forest. The pussy willow has been fixed to a hidden lump of Blu Tack to keep it straight. Celandine and primroses everywhere. Should we really be gilding the lily by adding lambs and horses from the farmyard set? I work hard at dissuading.

We spray the gardens to keep them damp and bring them inside to grace the table, ready for Clarissa and Zafar to see. Their train is due any minute.

Frogspawn

Georgie and I have seen frogspawn in the woods, a great heap of optimistic eggs laid in a shrinking puddle of mud. They have no chance. We take a family walk, over the slatted fence into the wood, not bothering to use the proper stile, and along the special thin path that seems private to us. Woo the cat is with us, slinking blackly along the top of the earth hedge and miaowing loudly. The larch trees are sprouting bundles of green, which look like a million dolls' house feather dusters, and despite earth works and tractor furrows into the path, some primroses are flowering.

We halt by the frogspawn. In the last few days no obvious tadpole development can be seen, but the host puddle is worryingly smaller and the edges of the grey jellied mass of eggs look dried out and shrivelled. It's been sunny for days and is likely to be dry for a bit longer. Hmm. We all crouch down and inspect more closely. The girls are afraid all the tadpoles will die.

Michael suddenly scoops up a big double handful of wobbly egg spawn and tells me to hold out his woolly sweater and he dumps the spawn into it. We look on with astonishment and then the girls join in excitedly loading him up with more dripping, slipping, gloopy handfuls of frogspawn. He is holding out his pouch like a pelican and then he rolls over the end of the sweater to contain the wobbly mess. He looks up with a triumphant grin and turns about to head home.

Frogspawn failure in the pond

Michael's deposit of grey, gooey frogspawn into the apology of a pond in the field has renewed our pond enthusiasm, and the girls have spent several hours with buckets of water, trying to make a better future for the frogspawn. It is floating in a muddy pool now and the girls are happy. So are the ducks. They came up from the river to see what we are doing and have paddled into the pond, very pleased with its better level of water.

They have also found the frogspawn and are guggling it down into their beaks. Should have guessed. Sorry, frogs.

Chicks

Ah! This morning we have chicks! The grumpy broody hen, who is snugged down at the end of the manger on a fine clutch of seventeen eggs, is making the most enviable maternal noises. I bend down to her level and see a small face peeping out from under her wing.

Now the mission is as follows: she needs to have a new nest area below the manger so that chicks falling off the ledge are not stuck, becoming prey to a rat or crow or magpie. At ground level, the mother can summon her chicks and huddle them under a wing. But if I move the nest today, I might compromise the future of any unhatched eggs; if I wait till tomorrow, then the early hatchlings might have fallen off the ledge.

What to do? Michael says to leave them till tomorrow, so there is the decision.

I long to know how many have hatched and what they look like, so I sneak my hand in under her wing and pull out a fluffy yellow chick and hold its ridiculous scratchy feet and small body in my cupped hand, letting its face peer out. The sharp egg tooth is still a white tip at the end of its beak and its eyes are black points. I return it and draw out another, and this one is brown with darker brown stripes. The mother gives me a cross look and pecks at my hand as I return her child. I have no idea how many there are, but I visit her two or three times to gather any fallen children.

When the girls come home from school, they rush into the shed and find the mother hen has made a decision of her own; she has jumped down off the ledge, ordering all infants to follow. Her abandoned nest is a clutter of discarded shells, none are left unopened, and she has hatched her whole clutch of seventeen babies. She has her work cut out ahead of her!

Michael and Childe Harold

I'm looking out of the kitchen windows to see if I can see Michael anywhere; he hasn't come back from picking lettuces. I peer through the low window where the fireplace once was and can see only the length of lawn and the fruit trees and the burst of flowers in the long bed. I stand on the yellow sofa and squat down to look out of the only windowpane out of nine that gives a clear picture and there he is, crouching at the edge of the ha-ha.

What is he doing? My curiosity is too great and, barefoot, I walk out to see.

He is still hunkered down and Childe Harold's great smoochy face is reaching up to him and Michael is stroking that sweet-smelling, brownly furred half-grown bullock. I stand and watch. Michael is pushing his fingers up and down that extended nose and tickling him, and Harold is trying to lick him, wanting to pull down a couple of fingers to suck on, great baby that he is.

We will have to take him to market soon. The harsh farming truth is that Childe Harold is prime beef, well-grown, meriting a government subsidy, and will make a fine butcher's carcass. Do we become faint-hearted and sell him at market, or do we send him to the abbatoir and bring back meat? These are troubling thoughts, which I push away; better to watch Michael and Childe Harold sharing love.

Guinea fowl

I am heavy hearted.

I am strimming the bank on the driveway, where the slate sign perched on a rock announces our farm name in drippy paint and my spinning, whirling line catches an egg, sending up a spray and splash of yolk. I cut off the fuel supply and push aside the uncut ferns to investigate and find a beautiful nest of twenty-three guinea fowl eggs, all cold, one mashed. The fox must have taken the mother. This is the price of independence.

A clutch must first be complete before brooding starts, so I hope these are waiting to begin and not that the tiny life in each of the eggs has started and then switched off, grown cold. I call Michael and we stand and look at them, first sorrowfully and then, with a futile need to mend, we gather up the cold eggs and put them into the incubator.

Hey, is this crazy? Remember, guinea fowl drive us mad.

They haven't hatched.

Tupperware

I feel elated, invited to a Tupperware party, definitely part of the real world. The venue is up the hill, buried in the new estate there, and it is crammed with parked cars. I feel suddenly very shy, not sure if I will know a soul, uncertain of the etiquette of a Tupperware party. I ring the bell and the door swings open to a wall of chatter and I don't recognise a single face.

Thankfully, no one seems to care or even notice me, so I lurk about and the hostess emerges from the kitchen with cups of tea, spots me and makes a vague general introduction. 'This is Rosie,' she says wrongly, and flutters her hand. The Tupperware lady has finished spreading out her wares and starts a little spiel about why they are good and what these containers are for. People are snatching them up and orders get scribbled on a pad. There's nothing I want.

The hostess leans towards me and says, 'Are you ready to order?' I feel panicky until I spot a large square box.

'Good,' I say, 'that might do for putting a leg of goat in to thaw from the freezer.'

'Oh no,' says the Tupperware lady, 'you can't do that, it's a cake box.'

'Yes,' I argue pointlessly, 'but it will be fine for the meat and stop the cat gnawing at it and dragging it about the kitchen.'

Suddenly things seem to have gone very quiet and the crowd of contented Tupperware buyers is looking at me, sensing trouble. The Tupperware lady is being firm.

'This is definitely a cake box and will be quite unsuitable for frozen meat.'

'Well, then,' I sigh, defeated, 'let me buy this box for cakes.'

It holds a joint of goat kid perfectly.

Chapter 10

Cutting the laurel

If we cut the laurel hedge that is growing high on a walled bank on the side of the vegetable garden, then the garden will get more sunlight. And the walkway to the end of the top triangle orchard will be more passable. But what a business it is. Either someone is on a ladder with a chainsaw or someone else is buried deep in the hedge trying to poke the saw in.

Michael once tyrannised an ex-boyfriend of mine by making him balance in the branches while chopping; it was an effective deterrent – he moved to South Africa. We have even paid someone on an hourly rate to battle with branches.

In each case it has involved much support from those below, dragging away the branches, lopping off side bits and making the stinky wood suitably sized to burn in the bonfire. Michael is in the laurels and the girls and I are below, wielding hatchets and axes to trim off the smaller branches. A four-hour stint gets about a fifth of the bank done.

As twilight approaches, we stop cutting and start the bonfire. It's the reward. I rummage in the log stash and bring out kindling pieces and a few good dry boughs to get the fire roaring, and then Morwenna and Georgie start to pile on the laurel branches. The leaves have a natural slick of oil in them and as we heap them on, they start popping and flare up with a fierce roar and flash.

We run to feed the fire with more. The wigwam of piled-up branches grows higher and higher and the flames rage up and start to singe the overhanging branches from the big trees above on the bank. The smoky steam carries up red and gold flecks amongst the ashy grey and we are covered by ashfall. It's a compelling, primitive craziness and we run around our bonfire, feeding it until the leaves and twigs are all gone and the middle of the bonfire has hollowed out.

Then it's our time to poke and rebuild the fire, pulling out charred, half-burned branches and feeding them into the middle. When there is no more prodding to be done, we leave, stinking of smoke, sad to part from this hot, red dragon we have made. Before bedtime, we come out and turn the smouldering ashes and it flares up again, to give us another surge of spark and red and gold.

The May gap

All bees have their June gap when most of their flower source is finished. We have a May gap for vegetables. Most of the seed packets recommend planting in May, but a few can be squeezed in early and so in May we have some seedlings, some not-yet-ready spinach, some not-yet-ready lettuce, and a number of straggling, half-rotten leeks left braving the spring rains after a long winter.

There is an ancient storage principle. It is that root vegetables can be stored in a clamp, where you make a circle of carrots and parsnips with their pointy ends facing the centre and you layer up with sand or peat, and then cover the whole stack with sand and help yourself over the winter.

Not so. Or we used the wrong sand, or made the wrong circle. Or the writer of the information was lying. Anyway, the clamp produces a tragedy of shrivelled vegetables sprouting hairy roots.

Budding saplings

We are walking down by the river looking for excitement and pulling away branches that create too much of an eddy in the water. Some ducks shout up at us. The poisonous hemlock is worryingly prevalent here, but the goats won't graze it. The buds on the alders are fat and purple and then I remember the pathetic copse I planted last winter, and so we sprint over to the sheep netting that contains it and make a leggy crossing over it.

'See if any are alive,' I suggest, and we are scattered as if on an Easter egg hunt, looking for signs of life.

'Here, here,' calls out Morwenna and we weave our way through the twiggy space and see she has found green buds. 'What tree is it?'

I stand back and ask Michael if he can remember what we planted. We compile a list: ash, rowan, oak, cherry, alder, sweet chestnut. We all peer at the line of buds, wondering which of these it could be.

'See what else there is,' I say, and we scatter again. Michael finds some nubby points. 'I've got an oak, I think,' he shouts, and we zoom in to investigate. Off again.

'Look, something like the river trees,' Georgie squeaks and yes, there are alder nubs. There are more and more, and we end up handling every twig and most seem to be assertively alive, offering little buds.

Those skinny striplings from the Woodland Project, so easy to plant in the dreary winter, so insignificant, so unlikely to live; a scraggy, bendy twig with a feeble drying-out root pushed into the ground, but hundreds of them. It seems like alchemy to thrust twigs in the ground and get a forest.

Amelia knows cars

Amelia decided to leave an old blue car in the log shed at Large Bottom Farm. It was an ancient Ford Popular . . . a good goer, sweet as a nut, that sort of thing, with leather seats and Morwenna and Georgie and all their friends loved it. It quickly became full of banana skins and dried ferns and small plastic dolls. The door sagged open and couldn't find the momentum to shut and then the tyres drooped deep into the wood shavings. Moss and other interesting sproutings took up residence, but the girls loved it steadfastly until one day when Amelia rang to say she had exchanged it for a bottle of red wine and someone came and removed it. I saw it driving through Liskeard.

Our red diesel Golf died on a day trip into Dartmoor. Car shopping is not an area of expertise to anyone we know, but Amelia is in Bristol, with heaps of oily fingered knowledge about cars. Together we approach a Toyota Corolla, and I look to her for the savvy actions of the cognescenti. We swagger forward, looking like the team you shouldn't mess with and she heaves up the bonnet. Inside is a shiny bright pile of car intestines that look as if they have just come off the production line, but Amelia gets technical, checking for clutch slippage and oil levels, then shrugs. I kick the tyres and she hisses at me, 'That's what numpties do.' We make a deal, hop in, drive home.

Archie helps with planting

Michael is in the vegetable garden with a cardboard box of seeds for planting: little gem lettuce, climbing French beans, spinach, broad beans, beetroot, leeks. The earth has been dug and we have spread well-rotted manure. The old Howard rotavator has been worked hard on this early bed, which is now a deep-brown, smooth tilth.

Michael has prepared his morning work by planting short twigs each width of the bed and running a length of baler twine between them, to give a planting line and to show where to watch for the emerging seedlings when hoeing out the weeds. He is kneeling on a feed sack and has a fold of paper holding the little lettuce seeds. With a pencil, he is carefully pushing the seeds out one by one into a neat groove in the earth.

At his side is Archie, our first cat. (We have a plan to do cats alphabetically, Archie, Bartholomew . . .) He likes this sort of slow, noise-free gardening very much and is stretched out on the earth, pretending to be asleep. Michael moves his sack along and Archie looks up with lazy eyes. When Michael moves the sack again, Archie gets up and joins Michael on the sack and watches the pencil work with interest. Then he wanders off to look at a bee. He hunches down a foot away from Michael and tenses up to pounce.

But it's not the bee he is after. With a flying leap he lands on Michael's back and aargh! – the pencil, the seeds, the box and Michael are sent spilling into the earth. When the lettuce seeds germinate, all the evidence is there: a line of seedlings and then, briefly, a chaos of sprouts, before the line returns to its proper order.

Juniper's babies

Juniper is making heavy weather of this pregnancy. Goats are usually matter-of-fact about the whole process, but no, Juniper is sighing, lying down and refusing to get up, flaring her nostrils and generally wanting us all to be sympathetic. She is like a furry brown ball with small, stick legs poking out under her and this week her udder has been swelling. I check the dates of mating in my Animal Movement Book and reckon that she is due any day. I leave her in the shed;

there is no point trying to tug at her collar and lead her to delicious grass.

By lunchtime, I visit her shed to see if she has moved out and find her nuzzling the most exquisite little brown baby; she has half cleaned it off and it is on the way to being fluffy. She murmurs a love song to it. I give her an encouraging rub behind the ears and check the hay to inspect the placenta, but can't find it yet, so leave them to bond, with the door shut to prevent other goats snooping on a private maternity ward. I will be returning in a short while with my post-birth treat of a pail of warm water with molasses in it.

When I return, I am not altogether surprised to find a second kid receiving love and licking; she was really huge and twins are commonplace. She breaks off her mothering and takes greedy gulps of the warm drink, before resuming her nuzzling. I leave the door open for her to refresh herself outside in the warm summer sun.

Later in the afternoon I visit her again, wanting to be sure of placentas; nothing is more annoying for a new mother than an infection from retained placentas, and I am astonished to find that Juniper is attending to a third baby. The other two are struggling drunkenly to their feet and colliding with each other, but Juniper is too busy with her newest child to push them towards her heavy udder. Morwenna and Georgie rush to the yard when they are back from school and pick up and stroke the baby goat kids and laugh at their feeble little bleats. They lead Juniper out into the field, each of us carrying one of her kids, and she settles to a mix of eating grass and kissing her newborn family.

Judy is the one who goes out an hour later to visit the happy pastoral scene and comes back to tell us that Juniper is licking a fourth baby.

It's extraordinary . . . four babies? What a feeding business that will be! We lead Juniper back to the shed and then take each baby in turn to check they can suckle, posting them at a teat and handling her low-slung, heavy udder at an angle to assist. She nuzzles at their backsides to encourage them and yes, each and every one is eager to drink but somewhat baffled by the teat, which hangs so low.

The eldest and strongest has the answer: she gets down on her knees below the udder and butts her head up, reaching for a teat and takes her fill. Juniper gazes into outer space and chews the cud.

Ducklings lost

A thin morning in June: Morwenna and I stand with our eyes squeezed up into the sun to look along the river valley. We are looking for ducklings. Their mother is desperate, knowing she has lost some, but is vague about how many she might have lost.

We have a sliver of time before school and climb up over the gate to walk along the river, hoping to find them paddling in an eddy. We both teeter on small grassy hummocks along the swampy patch with wild yellow irises and we push through the willow striplings to the water's edge. These upper reaches of the Looe river run too fast for ducklings.

We survey the dewy shine of the morning field. The sun is splitting through the oak tree on the bridge. I think a kingfisher darts past in a small blue flash. We can hear the mother duck's distant quacking as we find the first body, bedraggled, and then a second with slashed wounds. These are the unmistakeable killer marks of a heron. He will have eaten any others there were. We pick up the two wet bodies and sadly float them in the river down to the sea at Looe. We are quiet on our walk back home; Morwenna holds my hand.

Building a hut for goat kids

We have decided to move all the goats into the field by the drive; it's growing too high and needs some serious grazing, but the herd's four very new babies will need shelter if it gets cold or rainy, so we must build them a hut. I announce this very confidently, as if I build huts for goat kids all the time, and my father is roped in to assist; after all he has had a term of woodwork at an evening class. Michael goes to see what wood supplies he can find in the workshop and we reassemble in the field with a keen extra team of helpers in the form of goats and their skippity offspring.

The wood supply isn't much, and my father selects the best four pieces to be corners and we quickly see that the hut is only going to be mid-thigh high. 'It's okay,' I say, 'look at how little the kids are.' Michael mutters that we might as well give them an old tea chest and he slopes off. My father is unsure whether to cooperate with my gung-ho approach; he smells failure, but we have a lot of fun with banging in nails and holding pieces still against wobbles and collapse, and soon we have a box.

Georgie scoops up one of the babies that has sunk into a snooze in the grass and carries it up to the box, but it lurches onto little legs and potters out. Morwenna brings up a tranche of hay to make a snug den inside. We leave and then sneak back an hour later to see if there is any interest in our hut, and yes, inside are two white kids curled up together fast asleep in the hay.

I know that within a week those kids will be jumping on our handiwork and it will collapse about their ears. Shall I tell my father?

Pruning the vine

Our vine is the child of the venerable old vine at Hampton Court, the Black Hamburg, and it produces the most delectable black grapes in fat, tight bunches that sometimes are best eaten crammed into my mouth in a warm fistful. The others in my family are more delicate and Georgie makes a show of rejecting pips. It's supposed to need minimum maintenance, an essential feature for survival in our household, but of course, this isn't really true.

Apart from the traditional pruning on the shortest day, down to its three main fruit-bearing branches that stretch along the length of the conservatory, there is thinning, and harvesting and fighting my way through the copious amounts of greenery to reduce the jungle. I have a stepladder and secateurs and where there is a tiny form of a bunch appearing, I count down two more leaves and then cut. I have already made a series of selected cuttings of vine leaves suitable for making dolmades, but by now everyone is fed up with these.

The smell of the leaves is heavenly – green, sappy, juicy – and I gather great armfuls to load into a wheelbarrow and take down to the goats; at least they are appreciative.

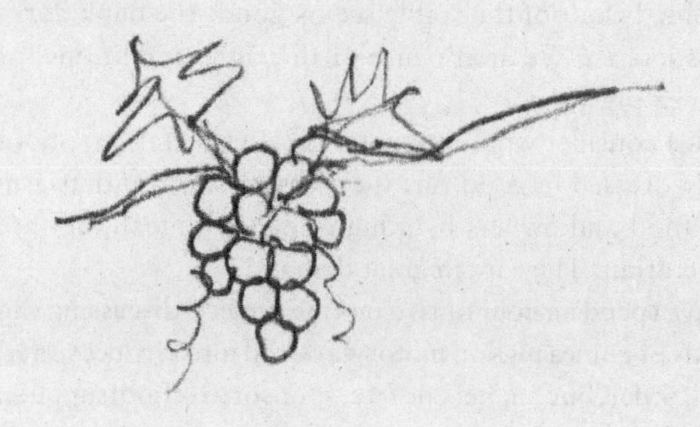

Tomatoes

In the greenhouse with the door slid shut behind me, all wind is left outside and the smell of ripeness is overwhelming. The aubergines and peppers are paltry things, but the tomatoes tower up into the beams of the greenhouse, pushing sprouting side branches through the slit of a window opening. They are crammed with fruit and I have come to pick.

Their smell is divine. I reach out to the low branches where the fruit is at its reddest and pop a tiny tomato into my mouth. It bursts into a riot of flavour.

I'm squatting on my haunches in a Cornish vegetable garden getting high on tomatoes.

Guinea pig project

Georgie longs for a guinea pig; we are happy to get her one, but where? There is a serious local shortage of guinea pigs. The pet shop apologises for a lack of supply and explains there is a great demand.

An idea forms. A serious demand and a serious shortage seem to be a business opportunity. I tell Michael that we will become guinea pig farmers. We stroll out to the big barn which is currently the only available space unless we squeeze pigs and geese together. The cobbled floor of the stable seems good, the dank darkness less so. We aren't sure of the right conditions for guinea pigs.

We consider what we know: they breed fast, are born fully dressed in good fur, they like to show enthusiasm for food and owners by a high pitched squeal, they are vegetarian. They are in great demand.

We spend an hour or two on this project, discussing van loads of guinea pigs on motorways and into city pet stores, mail order, buy one get one free, sponsored school supplies, until our fantasies erupt in laughter.

Kittens under the floorboards

Away for a few days, a small holiday. We are packing and rushing and shouting with excitement and rearranging piles on the kitchen floor, all ready for departure.

But we have a cat situation: Cordelia has five babies in a nest in the boiler house on an old sweater. Here she is safe and warm and dry. But today she is not there, and neither are her children. We put our heads together to think about where we have seen her. Yep, upstairs, by the second bathroom. We climb over each other in a rush to investigate and at the base of the towel shelves there is a hole for inspecting plumbing, a cat-sized hole.

We call her encouragingly and she pokes her head out for a caress.

Michael goes mental. We are told quite firmly that this afternoon we will be leaving and the cat will be outside the house, and if her kittens are inside, down a hole under some floorboards, well tough, that is the cat and kittens' problem. We are aghast.

The girls do pleading and wailing and I do reasonable consideration: kitten corpses under the floorboards will smell bad. He storms out.

I ring Duncan. He comes over and we lie along the floor listening for mewing. Cordelia is being kept away by Morwenna so the kittens will call and reveal their whereabouts. Faint mewling is tracked to the small bedroom under a chest of drawers. We pull up the carpet, wrench back the chest of drawers and lever up the right floorboard.

Five fat little bodies withdrawn from a dusty, gritty channel are returned to their far better nest in the boiler house, plus some extra cat stars for the grumpy mother. There is a passionate reunion and much purring. Michael isn't back yet.

Me, the farmer's wife

The girls are in the top barn playing, Michael is mowing and I am weeding the garden, when I spy a random child kicking around in the field. Not one of ours. He is about twelve or so and clearly has nothing much to do other than thrash some tall grasses with a stick. After a few minutes, he is close to the fence and so I straighten up and we look at each other.

'Hello?' I enquire.

'Are you the farmer's wife?'

'Er, yes, I suppose so,' I say, because it would seem a bit harsh to lecture feminist theory to this boy.

'Can I have some . . . milk? And a sandwich?'

I'm not really sure how to reply to this request, so I ask him if he has had lunch and yes, he has. I press on, 'So why are you hungry?'

At this he sees he is wrongfooted, so he blurts out, 'Well, you are a farm, and farmers' wives are supposed to have things, you know, for people who come round?'

'Well,' I say, 'you are more than welcome to come into the kitchen and have a glass of water and I will see what else I can offer.'

'Don't you have yogurt to sell?' he suggests.

'No.'

We traipse towards the house together, but he has lost heart. 'It's okay,' he says, 'I think I'm alright, thank you,' and he manages a dignified walk down the drive.

We are twenty years out of the loop. I should be plump and have a pinkish, round face and a stripy apron and my kitchen should be groaning with scones warm from the oven, waiting for jolly children to come frisking by.

Georgie and the guinea pig

Georgie is out foraging again. She has taken a doll's basket with a woven handle and lined it with dock leaves and is snipping shreds and tips of leaves from the bank into her basket. She is in the driveway, pressed into the bank and reaching up for a particularly select morsel of sorrel, and then she starts scissoring snips of primrose leaves, pennywort, and dandelion.

This devotion is for Peter, her guinea pig, a new acquisition. He is appreciative and gives a squeal of pleasure when he sees her and she thrusts these gastronomic delights his way. She says it is important to give him a balanced diet, and that he must not be allowed to get mouldy. We recognise these very words, which we have applied to her. So the message did sink in.

She is lofty in her responsibilities, going off with a determined air at assorted times of the day, returning with her snips. She clambers up to the table and gets going with a chopping board and sharp knife; cucumber, carrots and apple. She hums a happy tune as Peter shrieks at her from his cage.

The sound of music

Georgie has been given a violin. At a school assembly, a cool, big girl was playing. 'Does anyone want to learn the violin?' asked the peripatetic teacher and her little hand shot up and she was signed up. The lessons won't start until the autumn – meanwhile she has a violin and not the slightest idea how to use it, so she takes it out of the case and strokes it lovingly and makes hesitant small strumming noises on the strings. We are all concerned about the future of the violin and the violinist; there may be bad noises ahead.

It's a beautiful early summer day and Georgie is dressed in her favourite cornflower blue skirt, hair brushed and hanging long and blonde down her back; it's an unusually careful outfit and she has a look about her that makes me eye her suspiciously. She picks up the violin case and sneaks out.

I give her a few minutes and then move stealthily out behind her. I nip behind the wall when she pauses in the drive to pick a few campion flowers and then resume my tracking as she crosses into the haybarn, out along the narrow grassy strip, down the granite steps into the cottage courtyard and out into the field. I wait beside the big barn and move into the vegetable garden and lurk behind a row of climbing French beans and watch.

Georgie has moved mid-field, where the grass is quite long, and is walking back and forth singing her heart out and swinging the violin case in huge arcs. I get it straight away. The only possible violin work is to become Maria in

The Sound of Music and to be free out in the grassy slopes of the Cornish alps.

Egg in pocket

I am gathering up the feed bowls in the shed and letting out the animals. It is a beautiful Saturday with a feeling of unending weekendness; two whole days of time to spend. Tra la la, tra la la.

In the last shed I find a cosy corner occupied by a hen who has made a nest. I shoo her off, as there are more orthodox places for egg laying, but when she hops away I find an exquisite dark brown, speckled egg with a feather on it and, of course, it is warm to hold. I have too many feed bowls in my hands to manage an egg, too, so I slip it into the breast pocket of my old tweed jacket and I resume the clearing away in the yard.

On the wall, the late primroses are smelling heavenly and I pause to pick a bunch and then I think some other flowers would be nice as well, and I meander off to the top orchard for campion and bluebells, until I hear Michael in the drive. I run down to greet him, full of this joyous weekend feeling, and give him a bear hug, when I feel a strange pop in my chest. He feels it, too, pulls back and looks at me.

Is that a heart attack? No, it's an eggsplosion.

Rows of things

Michael is hoeing, Morwenna is weeding, I am weeding and Georgie is cutting delicate snips of greenery for her guinea pig. The sun beats down on our hunched-over backs and the earth smells rich and peaty. The weeds I pull up are fragrant, too. I lob armfuls of earthy groundsel and grass and newly hatching nettles over the fence for any greedy creatures that are snooping at our work, and the hens come rushing to take advantage of this bounty.

We are harmonious. I am on my knees scuffling under the long rows of beans in their bamboo lines, and Morwenna has moved to hoeing onions with Michael, following his instructions to avoid the guiding lengths of baler twine, which are held taut on small sticks to mark the rows. I am not trusted to weed any line that has baler twine, as I snag it and pull it up. At the end of my row, we consult and take stock of our huge vegetable patch.

The potatoes have sprouted and are in leaf, but no flowers on them yet to indicate readiness to harvest. Below them are eight rows of leeks, newly transplanted in holes way too big for their skinny state and then filled up with water, so they are lolling on their sides in a drunken disorder. At the end of the bed is sweetcorn, knee-high and looking good. The other bed has salad of four different kinds: rocket, cos lettuce, iceberg lettuce and sorrel.

The peas and mangetout are held up by a wonky arrangement of netting on bamboos and the mangetout are almost ready to pick. Beans, both broad and French climbing, onions, beetroot and then spinach and carrots. Every line is clean of weeds and looking like a garden catalogue.

In just a week or so, the weeds will have got the better of us and we will switch from precision weeding to crisis management. The picking will have begun.

Losing pigs

Two small pigs have gone missing. They are quite new and probably homesick. Or on an Adventure. There is a tell-tale hump in the fence netting in the yard where two small snouts may have pushed their way through, and who knows where they have gone. I give a cursory look in the field, knowing my search is futile. They could be anywhere.

I visit our nearest police station and request to report Missing Livestock. The duty officer looks at me blankly.

I attempt to justify myself, not missing Persons but Missing Livestock, escaped and therefore a danger. My complaint achieves only rolled eyes, but the officer does log it in the book. He spends an exaggerated amount of time on it, to show me how stupid I am: 'Names? Identifying features? Ages . . .?' They are new and currently nameless. Josh and Freddie? I leave quickly. I have pasta to make plus a million other essential things and I have wasted enough Pig Time.

Mid-afternoon, there is unexpected barking outside and then a jangling on our bell. A man announces that on a walk in the woods his dog has found some pigs, and might we know who is the owner? Amazing. We all join him walking down the hill and he motions for us to be quiet as we approach a large bush. He has a walking stick and parts some branches with it and we peer inside. Two small pigs are curled up together in a brotherly heap in a leafy nest.

We back off for strategic planning and thank the dog-walker profusely.

Michael and I now lunge in simultaneously and each grab a sleeping pig: a surprise ambush. There is ear-splitting squealing. We are not deterred, each keeping a very firm hold on an angry body as we march up the hill towards their pig shed. Gates are opened, the firm grip never lessens, and indignant Pig One and Pig Two are returned to their proper place.

We are both sorry for them and bring in an armful of fluffed-up hay, enough to completely cover them, at which point the squealing stops and we hear the much better sound of snuffling and new homemaking. We put a very heavy plank of wood down on the fence netting and resume ordinary life.

I drop into the police station the next day to announce the recovery of Missing Livestock. I am eyed suspiciously by a different Duty Officer, who looks in the record book and crosses me out.

Chapter 11

Making a patio

We are being powerful, effective women today. Michael has gone sailing and my sister Tamsin is staying for the weekend. When the cat is away . . . the big bad girls shall play.

Michael has for some reason shown no interest in expanding the small concrete strip around the house into a patio, on which one could dry onions or loll about, or even have a bench and lunch. Clearly this is something I must do on my own, but it's hard single-handed and if I have Tamsin here, the assistance of a sister is all that is required. And I have the girls if I can lure them in.

We take a professional look at the problem. We stake out the edge of the project and see that there is quite a slope to build up and most of our work for the day will be creating a bank, so we get the girls and ourselves and two wheelbarrows and go on a journey around the farm gathering good-sized bits of stone. Stan, the Dutch boy who is staying in one of the cottages, thinks this sounds good and he joins in, too. This will certainly keep Morwenna on task.

All deliveries of stone are tipped up onto the lawn under the big yew tree and we start by digging a foundation trench to lay down some big pieces. It is alarming to see how quickly a pile of stones can shrink, and how greedy a line of foundation can be. We send the chatty workforce out for further stone-gathering. Stones that thought themselves safely resting in perpetuity are yanked up or levered up with a pickaxe and we find a few blocks of granite lurking under a hedge.

We have a chipboard square on which to mix cement and we find ourselves very taken with this project, some on stone selection and stone placement, and others on cement mixing or cement application with a trowel. Tamsin is in charge of the spirit level, I'm back-filling with earth, Stan is positioning slabs and stones, and Morwenna and Georgie are passing choice pieces to him. Then the wall builders and cementers become earth carriers and rubble collectors.

The day spreads out into bursts of wall building with a great deal of wheelbarrowing.

When Tamsin leaves, I start to think defensively about what to say to Michael. I need his help with the screed on the top.

The paddling pool

The red paddling pool is half-full, the children are at school, the day is hot. I have been working hard in the garden pulling up weeds from the base of the climbing beans. I have stripped off to my pants to do this work and I can feel my skin stretching tight in the sunshine over my shoulder blades. Eventually, I gather up the mess of ruptured plants from where I have trod on them and I bundle them into the compost box.

I throw my clothes over my arm and pick up my shoes and tread gingerly on the stone path up to the house. I slump on the slate box to rest. Stupid plan, the slate is nearly red hot. The red paddling pool is shiny, and the water has the slightest wrinkle as I stir it with my finger. I strip off my remaining garment and step in, so cool, so cool, and I stretch out and let the morning's labour drift from me.

Paddling pools should never be the preserve of children alone, who have been playing with the sturdy plastic slide, climbing up and slipping down into the pool with a splash.

I step onto the top of the slide and start a descent. Wonderful. Up and down I go and then start on the variations. I lie down and start a journey head-first. Only a few inches from the bottom, the slide narrows or I widen, and I am suddenly and painfully jammed in by my hips.

I am wedged, naked, upside down on the children's slide in the paddling pool. When the agony has subsided, I twist and push myself into reverse enough to tip myself off the slide and fall back into the pool to recover.

I have cooled down, but even before the children are back from school, I have livid purple marks on my hips.

Horrible red

I have a plan for today's pickings of a few random blackcurrants, a heft of blackberries and five raspberries. I shall make a pudding we loved as children, made by my mother from redcurrants. She called it Rodgrod, which of course defied any argument. What on earth could it be? But whatever it was, it was delicious and we loved it.

I phoned her for a recipe and was told to simmer my fruit, did I have redcurrants? 'Yes, yes, of course,' I lied. I must simmer it, drain and sieve it and then add to the thick fruity juice two tablespoons of cornflour and sugar to taste. Easy. Then I must stir it over a constant and gentle heat until it thickens, no lumps allowed, and sprinkle a thin crust of sugar on the top and allow it to cool.

I serve up a livid carmine bowl onto the supper table. Elbows are on the table, spoons are held aloft, and Georgie looks at my fancy swirl as I place it on the table and peers into the bowl. 'What is that horrible red?' she asks.

'You are horrible red,' I reply smartly and slop a spoonful into a bowl for her and serve in cordial gentility a spoonful or two for Morwenna and slide back into the shadows.

Morwenna dives in and declares it delicious, how I love that girl. Then Georgie has a pouty-mouthed, teeny little taste and she too finds it delicious. So starts a famous family pudding not always made with redcurrants and forever known as Horrible Red.

Geese in the orchard

We have briefly become drovers, a pleasingly timeless occupation, driving our geese down the drive to the old orchard. For our geese, it's just as exciting and terrifying as the serious, ancient driving between towns with tar paddled on their feet for a journey of a hundred miles. They are alert and panic stricken, yet trust Michael, their beloved leader. They follow him, but look from side to side and worry why I am bringing up the rear.

Michael has opened the five-bar gate and welcomes them in with a flourish and I shut it behind them. They have the rickety goat box we made as night shelter – not much protection against foxes, but this is a short-term experiment to see if the geese are good lawn-mowers in the orchard. It's a great idea to get geese grazing between the fruit bushes, where it's so difficult to push the mower.

As it turns out, it's not a great idea at all, and we crossly drove and drive them all back up to the farmyard. They have eaten the gooseberries that were hanging almost ready for picking. I am snorting with annoyance. Later Michele looks at me in her quizzical French way and says, 'But of course, they are geese, *non*?'

Geese, gooseberry?

Grace Periwinkle at large

Morwenna says Grace Periwinkle, her rabbit, has decided that she needs the full run of the garden. Morwenna has accepted this and maintains that her rabbit knows and loves her and will always come home. This seems to be true for the first days, but now comes the time when one large and wilful rabbit does not come home to her tender-hearted owner and the cage in the garden is left open in case she creeps back in the night. There is anxious talk of foxes. And days pass with no sighting of Grace Periwinkle.

Finally the rabbit saunters back home and is decidedly off-hand with her loving owner, picks nonchalantly at a feast worthy of the Prodigal Son, and avoids being caged for the night by hiding in the impenetrable rose thicket. Morwenna decides that if we can catch her, she needs to go back into the cage forever. Grace Periwinkle views us suspiciously from under the rose thicket and isn't coming out, so there she stays.

Summer croquet

The old wooden box set of Jaques croquet. Four mallets, each with a different coloured stripe on the handle. The mallets are showing signs of age and fatigue, or maybe they have been used for bashing other things. Four balls, round, heavy and ridged, scarred in places, and in blue, red, yellow and black. Six hoops, a central pole for pegging out. The hinged box lives in the conservatory and is dragged out frequently; it's a sign of summer. The lawn is mowed and cunningly disguises the holes and dents made by invasions of badly-behaved bullocks. Only we and our regular opponents know where the holes are.

The kitchen window is open enough for music to trickle through. We have glasses of Pimm's and beer. The women are wearing frocks, even me, especially me. It is early evening and the girls are scampering about and can put themselves to bed later. Morwenna crowds round Michael, willing him to make a crack shot, and he obliges. He is glad not to be partnering me, as I am too reliable a bungler, and I flit around the lawn feeling elated and summery.

The end of Grace Periwinkle

Grace Periwinkle staggers down the granite steps as if she has had a thoroughly wild night on the town and I quickly call Morwenna. There is a brief moment of girl/rabbit reunion before Morwenna leaps back and says there is something wrong with Grace Periwinkle, who we see is trembling and hunched with her fur looking prickly. Oozing from her eyes is filmy liquid. Her eyes are swollen, bulging even.

Is it possible that here in the woods she could have contracted myxomatosis? From encounters of a wild kind with other rabbits?

We tell Morwenna what we think, that her rabbit is likely to die miserably, and the kindest thing we can do is take her to the vet to be put down. Morwenna, set for school in her neat uniform, readies herself and puts Grace Periwinkle in a cardboard box with some wisps of hay and a few pulls of grass.

'I don't feel bad,' she says, as we drive to the vet. 'Grace Periwinkle had a much nicer life being a free rabbit.'

Horses

We have stables, we have hay, we have grassy meadows and a river, we have two girls. The rest is obvious. We are to have horses. Michael huffs in a disapproving way. He was born with horses, rode them all through his childhood, and in my view, he knows horses. He had a nifty little showjumper and a bad-tempered pony called Jimmy. Oh, he knows. But he says gloomily that all the time he hated it, hated riding and longed to be sailing.

I once watched him rescuing our neighbour Pippa. She had a crazy horse galloping round the field refusing to be caught and Michael set his mouth into a firm line and took charge. With a bowl of oats in his hand, he stood as still as the gatepost and looked nonchalantly away. The lure worked and he threw both arms around the horse's neck, locked his hands together and off they went careering round the field with Michael absolutely certain that he wasn't going to be the first to give up. After several laps around the field, the crazy one quietened and Michael led him into the stable.

The girls have asked for riding lessons. Michael puts on a grim face about this. He has decided that we will never initiate anything horsey, but if they beg, harass, hassle and torment us about having a pony and the begging lasts for at least two months, then we will consider it. Two months have passed and it seems that they have forgotten about horses altogether.

Michael is packing away his horse skills with a smug look and is buying sailing magazines. I am alarmed.

Apple picking

Apple picking is another of those team efforts. One person up the tree, perched with a foothold and leaning far out to the fruits one cannot reach from the ground. Another person stands below, ready to catch. Others might be persuaded to join in on other trees, or with a bucket for windfalls. One might serve as a lookout, shouting out the position of ripe or large fruits. On this particular day, there are only two volunteers.

Michael is up the tree we called the Leaning Tree of Pisa. The ladder probably isn't really necessary, some careful climbing can do just as well, and then maybe he wouldn't slip. Up the ladder again, and still yowling about his painful ankle, he manages well enough to get to a clutch of apples and snag them off one by one, with stalk still attached, and drop them down to me. I manage the catching pretty well. I lay each gently in the box, careful not to bruise them. These apples are a good variety and will last the winter in the top barn.

Once he has cleared the lower branches, Michael edges further up into the tree and I can't see him anymore. I call out, 'Let me know when you're ready and I'll get them.' No sooner said than an apple comes hurtling down and hits me on the shoulder.

'Hey, not yet!' Another apple breaks through the leaves and branches above and hits the ground. 'Wait, wait, wait, I don't know where you are. I'm not ready.'

Things aren't going so well. Michael yells, 'It's obvious where I am, idiot, above where the apples are falling!'

'Just say when you are going to drop.' One comes down immediately and I catch it. 'SAY when you are going to.' Another apple. I still get it, but I miss the next one and the one after hits me on the neck. 'Why don't you TELL ME instead of hitting me on the head?' I shout this time.

'Because you should be there, watching, while I risk my life up here.'

The apple picking, catching, dropping, and shouting continues like this. The man in the tree is an imbecile.

I really ought to start throwing them back up at him, but I am too lousy a shot, he has the advantage, and besides my neck and shoulder are hurting. I start yelling abuse at Michael and then stomp off. This is a victory for me, as there is no point picking apples without someone to collect them; so down he comes, in for a cup of tea, grumping, and eventually he apologises and we start again.

It's a beautiful harvest in the end. We take one side of the box each and haul them up the vertical ladder to the top barn, where we space them out on sheets of newspaper. The whole barn smells appley right up to the rafters, and looks like a harvest festival. We put the short-term apples nearest the ladder, those Grenadier that are so good for cooking but don't last. The eaters are in the middle, a smaller crop of James Grieve, but huge in size; last go the Mare Tree, the Leaning Tree and a few Bramleys, our winter store.

The next day in the town, I see our old neighbour, Ernest the vicar, from across the valley. 'Good morning, my dear,' he says. 'I heard you having a spot of bother with the apples yesterday.'

Nutmeg eats a chicken

We are hosting a family gathering, somewhat unplanned and a very odd mix. Michael's parents are staying, his sister too, when my brother Peter arrives to camp in the garden with Jane and their two little girls. It's fine. I like the lawn as campsite and I like everyone bumbling around.

I feel it's helpful to give out a few ground rules when I see that not only do we have the parents' poodle, but Peter has also brought his famously ill-disciplined Jack Russell terrier, Nutmeg. 'Nutmeg must be on a lead until she is under control,' I insist, but get the quick answer from Peter that it won't be necessary, she will be fine. She disappears. Immediately.

Everyone is busy saying hello and bringing out stuff from the car and scampering around and being offered tea and cake and then we hear it. There is a hullabaloo in the yard, ducks and geese in full voice. I look at Peter and he stretches in a leisurely way and says, 'Let's see what's going on.'

We don't get as far as the yard before Nutmeg streaks back into the garden with a chicken in her mouth, which is squawking and hollering. She races around the garden as if it's rodeo time; none of us can catch her and she's loving it.

'NUTMEG!' yells Peter and makes a failed rugby tackle for her, while Michael's confused father looks on and asks, 'What is the problem?'

'Bloody Nutmeg!' I shout at him, as I run past, trying to reach her.

Peter is up on his feet again and manages to grab her and together we try to wrench the hen out of her gripped jaw, but she is clenched tight. At last Peter wallops her and we retrieve the chicken. Dead.

The smell of buddleia

Buddleia attracts butterflies and me. I see them now fluttering around the two bushes, a standard, pale mauve buddleia that grows in every wasteland site, and our garden specimen in deep purple. There are tortoiseshell butterflies and cabbage white and a few others I can't identify. They are swooning, I am sure, with the heady scent of the plant. It makes me swoon, too.

I am trying not to frighten off the butterflies, but I have my nose buried into one of the marginal flowers, whose point is drooping and not yet blossoming and whose swollen body is a mass of tiny flowers with a coloured inset.

The smell is . . . summer, laziness, childhood.

Onions

More than a hundred and fifty onions are crumpled, stretched out, curled up or sleeping in the sunshine outside the conservatory door, resting for the third day to dry out their fleshy necks ready for winter storage. It's a fine harvest and none of us mind stepping round their indolence.

We turn them, taking care that the white roots are turned up to shrivel and dry and the long messy necks are folded over so they too crisp up in the sunshine – with any luck turning to threads of raffia, so that a great bouquet of onions can be plaited together and hung from the beams in the big barn. Onions with necks too short get laid out on corrugated sheets on the floor in the big barn alongside the apples, combining to make an acrid, cider smell later in the winter air.

Morwenna and I kneel beside the onion pool and reach out and turn each one, trying to keep track of which have been turned. The skins feel dry and rustle in the hand, any earth has already fallen loose. The cats step through the onions, patting and turning them where they shouldn't. A wheelbarrow is waiting close by to trundle away any dried-out onions ready for storage and as the days of sunshine creep by, the pool gets smaller. A trug is at hand for any onions showing soggy or mouldy patches, to be made quickly into onion-rich ratatouille and maybe French onion soup.

Finally, the last onions have found their place and all that is left on the warm concrete pathway is a tawny yellow-and-ochre litter of onion-skin fragments and crumbs of dried earth.

We need to get rid of those trees

Behind our farmhouse is a line of mature pine trees. If you stand down by the river and look back, the trees form a backdrop behind the house. They capture the west evening sunlight in their feathery fingers and throw down little shards of evening sun and then the house is cast into gloom. We have the glory of sunrise, some sun in the day, by afternoon the hip of the house and the conservatory has a little sunshine, but the evening sun I covet is the sun we cannot have.

How can I ask a neighbour to cut down a stand of trees? The answer is that I can't. So we gobble up pourings of evening sun in the vegetable garden and we have small picnics chasing the sun as it slips away towards the river.

The house is hunkered down under the hill, snug and safe from the westerly winds that prevail. This week that westerly wind is strong enough to tangle and rip up one of those big old pines and before I can properly hope, before I can even dance and gloat, Jack and Colin and the helping man are in the copse with chainsaws. All of the pines are taken down and suddenly we stand in a pool of sunlight.

Swearing box

Georgie is outraged that there is swearing in the house.

Swearing for Michael involves shouting or muttering, 'F—ing Ada.' We are cautious when Ada is mentioned. Even Georgie. Now she has initiated a swearing box in the form of a jam jar and a written tariff at its side. The cheapest swear is to say 'bloody' and that costs the offender five pence. At the other end of the scale, to say 'fuck' will set you back fifty pence even if attached to Ada. Georgie has a small fund growing steadily and says it will be for buying holiday ice creams.

Philip arrives with his family and trawls around the house picking things up and fiddling in the time-honoured way of seven-year-olds. He spies the swearing box and studies it carefully. He correctly reads the tariff and has an epiphany. 'I've got fifty pence!' he cries. He fishes out his fifty-pence piece and unscrews the jam jar.

In a moment of careful stagecraft, he faces his audience and slowly and deliberately drops in the coin. 'Fuck!'

A swarm

One of the hives has swarmed and we must set up a new hive before we collect the swarm and bring it home. It sounds straightforward, but it is full of difficulties. First find your swarm. The bees are helpful with this; they set up such a smooth hum that it can be heard if you are close enough and we track the sound to the high branches of a sycamore tree. Preparations. Two people protected by bee suits and hats and veils and gloves. A ladder. A cardboard box carrying the tantalising and apparently irresistible smell of some blackcurrant leaves. The smoker. A small bow saw.

I'm nervous as I hold the ladder steady. There is too much stuff to take up. Michael climbs to see if he can reach – only just, and comes back down. He thinks he doesn't need the smoker. He slowly ascends with the box and the saw jostled between the ladder and his chest, and positions himself below the swarm ready to saw through the branch onto which the bees cling in a rounded cluster. He drops the saw, and I dodge it as it hurtles past. 'Michael,' I hiss, 'what are you doing?' I strain to look up and see. 'You just keep holding that ladder still,' he hisses back. He shifts the box around and then gives one deft smack on the branch. The swarm drops in the gaping mouth of the box. We have our swarm.

Fuzzy glass

Our east-facing window has nine panes of glass. They are the original, put in when the manor farm was burnt down and the site became two conjoined workers' cottages. Later the land was parcelled up and the Duchy of Cornwall granted a farming lease with sixty acres and the cottages were turned into a single dwelling. Someone, sometime, smashed a pane and it is only through this single pane that reality is shown. The other eight panes are slurred and have thin lines that blur, like the mix of thick gin in tonic.

Standing in the kitchen and gazing out is like studying a Monet. The high trees on the right give a brush-stroked range of greens and the little white house opposite is hatched greys and whites, with an echo of pink by the door, and wobbly patches of browns and other greens where Ernest's vegetable garden and serried ranks of daffodils lie. The sky in the chosen mood of the day is blue or grey and mottled.

So to stand and gaze is to look at our own expressionist painting and when the need is there, well, you must kneel on the sofa, and crouching, look through the bottom left pane to see Ernest and May weeding, or Georgie and Morwenna playing down by the river, or a line of ducks moving purposefully across the meadow.

Bella the goat

Dammit, the goats have managed to find a way out somehow and are in the old orchard, happily munching away at hedgerow and brambles. They shouldn't be there, but they aren't making trouble, so I decide to be lazy and leave them for now.

The next time I look out, they are making a lyrical picture, lying about on the concrete by the front door having a rest and cud chewing. This time I think I ought to push them back into the yard, but they are sprawled so sweetly. I am so beguiled by them that I fail in Risk Assessment. Close by in the hedge that separates the orchard from the lawn are some poisonous plants.

My next lookout is on a corpse. Bella, the only goat we have purchased in recent years, a goat of excessive stupidity and high pedigree, is lying dead outside the front door. I am completely astounded.

Michael and I discuss the likely cause of death and think it is heart failure brought on by eating poisonous leaves. Why are they not all dead? Because most goats do not choose to eat poisonous leaves.

I check out the goat shed before I go to bed to see if we have a herd living or dead, and a collection of goats look up at me with those enigmatic yellow eyes, much too wise to be felled by dodgy leaves.

A bee problem

It seems a good day for inspecting the beehives. Michael at work, kids at school, sunny day, washing all done and as it's June, the starvation time for bees, I should see how many supers are full and what emergency rations are needed. There is a strong hive in the old orchard and two weaker ones up on the top triangle field, so I will be busy.

I dress properly. I have a layer of loose clothing and then the strong, white cotton boiler suit and the leather gauntlets, the wellington boots with the trouser legs over the tops, and the bowler hat with the gauze net. I light the bee smoker with some newspaper, find the hive tool to prise apart the wooden box sections, the supers, and start off from the conservatory, when I decide that I am so hot that I will be pouring sweat and die of exhaustion before I get to the hive. So I stomp back into the conservatory to strip off the undergarments.

It never pays to be a hot or stressful beekeeper. I am no longer in the same lovely mellow mood as I was before and probably the bees sense it, because when I open up the first hive and take off a super of honey to look in the brood chamber, the bees are making a fuss. There is a general hum of dissatisfaction from them as I check for the queen, a lovely circle of eggs, some well-provisioned cells and supply of honey.

I am about to restore the hive to its proper order when a gang of distinctly troubled bees, only a few in number, decide that I am an intruder and should be chased off. I get three or four stings in my gloves, which don't hurt me, but this sort of sacrifice from their fellows make the rest of the bees anxious.

I quickly put the hive together, but not before a large number of bees decide to buzz angrily around me. I brush some off my cotton suit and sense that a few have got into the collar, no, not a few, more than a few. I must get them out or take off my suit very quickly, or I am going to be badly stung.

I walk steadily towards the gate, keeping calm, or trying to keep calm, and by the time I get there, I have three stings in my neck. Just ahead of me is the narrow entrance to the wood, dark and shady and unloved by bees, so with clenched teeth and deep breathing, I tear off my hat, throw off the gloves, unzip the suit, step out of the wellies and run stark naked into the wood and run, run, run, until bees are no longer following me and I am safe.

How long should I wait, naked in the wood, I wonder, for a safe moment to return home?

A plan to kill the fox

There has been strong talk about foxes. Colin says he has seen several as he travels around his farm and they have mange, their tails are thin and they are desperate for food. I talk to Michael about killing our own marauding fox.

It is almost full moon and the weather is set clear, so we can have adequate light all night long.

Phee says she will sleep over and we set up a comfortable watching zone in the big barn, where we can view out of the unglazed slit windows. We drag a mattress over and some cushions and bring out the shotgun and a box of cartridges. We have torches, a bag of dried figs and warm jumpers. It is to be a night-long vigil, taking it in turns to crouch by the slit window and watch with the gun readied at our side.

We creep out at ten and feel giddy with the excitement of the plan. The barn creaks and groans with our footsteps on the staircase and on the long flooring planks. We each take a quick look to see what visibility we have and the long view over the moonlit fields is breathtaking. The oak in the distance is silhouetted, and the slope of the long field with the cows moving slowly as they graze is all in shades of misty purple. The river glints. Our geese and ducks make the occasional sound, but otherwise it's quiet, and we wait. And wait. And shift carefully on bent knees, and swap almost silently, and wait, and eat a fig, and wait. There is no sight at all of a fox, a rabbit, a weasel, a badger. Nothing. On my two o'clock watch, for an agreed hour, Michael curls up and sleeps. I wake with my head in an impossible angle, neck aching, Michael still asleep and dawn breaking. Oh well. We slink back into the house and crawl into bed for an hour or so.

Chapter 12

Arrows

Are children like arrows? I have drawn back my bow and hurtled Morwenna far. She has to take a train to her secondary school, weighed down by bags, and scamper across a park to the school. Dark, late, on her own or struggling to keep up with some big girls.

War zone

I can't bear it. For weeks now I know there has been big forestry work going on in the woods; tractor marks first, then a gradual piling up of trunks, then the forest pathway totally obscured in mud and deep tracks. I can still walk there if I take one of the secret paths, but the walk itself is robbed of pleasure. I am looking at a battlefield.

As the ground falls away into the river, the whole slope is a mess of tree stumps and hacked-off minor branches and random sites of bonfires still smoking from charred remains. I think of Macduff and the horrifying news of his family, slaughtered. That's what it is here, slaughter and destruction.

Before I turn back, I see untimely primroses crushed in the mud and wonder if I should be here with a trowel ready to transplant. I'm fazed by the devastation of this wood, my beloved wood.

I take a deep breath and remember how all those years ago the larches were felled and replanted, and how the clearing of one area somehow opened up another; how the wood keeps changing as some parts become almost impermeable. I think that the ferns and the primroses and the brambles will ignore all this man-made bother, and I shall wait for spring.

Railway daffodils

Down by the bridge and railway crossing there are two allotments, one managed in a ramshackle way by Mr Warboy, and the other long overgrown. If we creep through Mr Warboy's and into the next, there are daffodils and snowdrops in profusion. What is more pleasing than gathering wild things? We pick daffodils from the dense jungle of overgrowth; they are being strangled by briars and we are intrepid foragers.

I pause by the bank where earlier a storm of snowdrops had blazed. Now they have swollen seed heads and wilted petals. It is at this point that they can be transplanted. If we leave them, the aggressive brambles will overcome them.

I am firm in my belief that we would be performing a citizen's duty in digging them up and transplanting them to a more loving home in our orchard. I spend most of the journey up the hill trying to persuade Michael of the rightness of my plan, but he calls me a countryside vandal and leaves us girls to pillage on our own. We take down trowels and two plastic bags and wear gloves as defence against brambles, but carry within us a troubled conscience.

'We will be saving snowdrops,' I tell the girls. So we dig.

Locked in the goat shed

Everyone is tucked up in bed; that is, children and animals. Michael and I are at the early end of an evening. The kitchen is cleared and he is marking schoolbooks in the study area. I announce that I am slipping out to check on the goats, as Yamaha is heavily pregnant and needs regular checking.

I put on an old tweed jacket and my boots. I don't even bother with a torch; it's dark, but I know my way and besides I love the way the thin moonlight distorts the walls and hedges, turning them into spectres and huddled figures.

It's quiet as I step softly along the gravel. I'm trying to creep into the yard without waking up all the geese and ducks. I'm almost successful, just a few half-hearted sleepy quacks, and I lift the wooden latch of the goat-shed door and slip in.

In the first shed there is a rather charming little row of goats sleeping on the wide wooden shelf on the manger. They have tidy feet tucked under white bodies and seem like a choir looking over the congregation below. The congregation is limited to two goats pressed close to each other, who do not move or acknowledge my presence. I leave this shed and let the latch fall shut and I move to the next shed.

Here Yamaha is standing up and greets me with a snicker of sound. Plum is asleep in the straw. I stroke Yamaha's nose and slide my hand over her taut belly. She is uncomfortable for sure and I can feel her babies giving out sharp kicks. Perhaps she will start labour tomorrow, but for now things are okay.

I turn to leave and push the door bolt, which should slide to one side to let me out. It doesn't budge and I realise that I have not flicked back the wooden latch on the outside and it has fallen down and locked me in. A thousand responses flash before me: that sickening feeling of being a complete idiot, to rage, to futility, and then tiredness. I gather my wits and think.

The visitors staying in Owl Cottage are near enough to hear me if I shout, so I start up a yelling and find a piece of slate to bang on the door. The ducks and geese quickly join in the din and I wait for rescue. When nothing happens, I decide that the only rescue will be Michael, who will soon be ready for bed and puzzled to know where I am. Actually, I don't have a clue what time it is. Please let it be bedtime.

I go over to where Plum is curled up and sit down beside her in the straw, stroking her warm, curved backbone. Yamaha comes over and breathes a whiskery breath on me and snickers again. It's comfortable enough, so I lie down beside Plum with my head on my hands, and a bit of sacking and some straw hunched up on my back, and I listen to the night noises until I fall asleep.

Minutes? Hours? I am woken by the noise of the bolt and Michael shines a torch into the shed and finds me in my huddle of straw. 'What on earth are you doing here?' he says.

I shake myself up and ask, 'What time is it?' He tells me it's half-past midnight.

I have been in the shed for hours and I'm sleepy and stiff, but I gather up fury and storm out. He follows, fast to keep up with my super speedwalk. 'What? What were you doing?' I do not reply.

'Hey, are you annoyed with me? Wait, I haven't done anything wrong.' I still do not reply.

'If you locked yourself in the shed, it's not MY fault.'

I turn to face him, square. 'No, it's not your fault that I was stupid enough to lock myself in the goat shed, but it's your fault that it has taken you HOURS to even notice that I was missing.'

Michael can be heard, annoyingly, chuckling on his way back home.

Visit from the herd inspector

Aargh, panic stations. I hear a car rumble up the drive and a man in a dark suit with a Barbour jacket over it has stepped out and is coming steadily towards the house. He looks official and I scan my brain for who this could be, and have a flash of premonition that it could be the man from the Ministry of Agriculture, coming to look at my Animal Movement book. I am entirely right and totally unprepared.

I greet him warmly and invite him in for coffee and fill the air with farming talk and chat about the state of the market. At the same time, I am trying to calculate how up-to-date my Animal Movement book might be and how many movements there have been. I hope he won't ask to look at our ear-tagging system, which is a little crude; those marker blobs on a white ear.

I think that two goats have been slaughtered and one goat sold in recent months and not much else, but the moment comes and I have to reach up into the bookcase for the scruffy red exercise book.

I see I have not entered those three movements and so I say brightly, 'Oh, bother, I didn't put in the last three, let me do that quickly now.' And I grab a pen and write out the details. The man from the Ministry is a nice guy and knows he has caught me out, but I am the smallest cog in the very big wheel of agriculture and three goats are just three goats.

He signs the book with a flourish and I wave at him as he leaves.

Egg blowing

In my basket I have two goose eggs, five duck eggs and some hen eggs, all have been carefully chosen for artistic value. The goose eggs are enormous, chalky and would be white if I washed off the muddied look. Duck eggs are coated with a fabulous greeny blue veneer like a thin lick of paint. Hen eggs come in brown, brown spotty, brown freckled, dark brown. I have an artist's palette of egg.

They are going to be blown for painting and hanging on an Easter bough.

It's a gruesome process. We need only the shell, so the yolky insides need to be flushed out. 'Make a hole each end and blow the goo out,' says my father, a childhood collector of birds' eggs. Over the years we have learnt that spiking one end with a needle won't do it; the force needed usually cracks the egg. Michael has developed a technique with a drill.

We assemble egg boxes and Michael drills first one end and then the other. I then poke a long needle into the egg through the hole and try to break the yolk. Then we have the kiss. Pucker up and press the lips to the eggy hole and blow firmly, holding the egg over a bowl. 'Yuck!' cry the blowers. Gloop sludges out of the hole in a reluctant glop or a runny rush till it's empty.

With a straw, try to let in some water to rinse the egg out, or if it's a greeny duck egg whose veneer needs protecting, maybe just leave it. Let the eggs dry out. Assemble an array of paints, colouring pencils, sequins, nail varnish, newsprint, glue, feathers, torn shreds of poetry, and start decorating.

We have a keen gathering of painters and decorators, vying with each other for artistic inspiration. Georgie is going for total green, Morwenna has made a map of the world on a goose egg, I am over-ambitious and have almost completed a cherub with wings before I press too hard and have taken a chunk out of her wing tips, and Michael has created what is supposed to be a chick in yellow with legs like tree trunks. We hang the eggs on a branch.

Someone has left both the kitchen door and the front door open. The eggs are moving very slightly in the breeze, but then a gust swoops through the room and fells the entire Easter branch with a smash on the floor. We cluster round in dismay, trying to rescue where we can.

Culvert in the copse

Georgie and I have pootled down to the copse by the bridge. We stroke and admire the buds and leaves that have appeared on the twiggy saplings and wonder how tall and foresty the copse will become, maybe harbouring wild creatures. We stray to the water's edge, where the river has sagged the bank and made a marshy pool. We wade about, squelching in the mud and count how many wild irises we can see.

I tell Georgie, 'If we are lucky, we might see a kingfisher swoop under the bridge,' but then I feel a cheat, because I suspect that will never happen. Georgie settles down to wait for one. Eventually, to distract her, I suggest we pick the remaining daffodils that are in the copse.

Halfway along the copse, my foot makes a hollow noise and we scrape away the turf to see a flat piece of slate. It lies alongside another. We scrape some more. It seems that there is a pathway of hollow-sounding slabs. The fourth slab has a pile of earth and tangled roots, which we tug and finally haul up, and in its place we see that we have uncovered a culvert, a drain, made of slabs of slate laid like a square channel.

We are very excited by this discovery, and decide to gather more tools, gloves and better still, brag about our adventure. Soon we are an archaeology team, hard at work clearing this new-found culvert, and at fifteen slabs cleared, we notice a seeping wet mud forming over the base; it really works, this drain is bringing water from one of the many springs that rise in the hillside. We clear the exit as it nears the river and tread down a final channel with our boots. Maybe we are on the way to clearing the squelching marsh that sogs this part of the field.

Goodbye to bees

I have found a novice beekeeper eager for our hives and the equipment. Spring is a good time to transport them and settle them in a new place for a busy season of foraging. But we are so sad to let them go. We like a murmuring summer morning browsing through the hive, spotting the queen, checking for queen cells that might mean a swarm is brewing up, looking at the exquisite patterns in the hive of hexagonal cells filled with goodness.

Last summer I found some cells filled with black treacle and puzzled about it until I realised it was road tar that had been lying sticky on the ground. Some cells were packed in with threads from my red mohair sweater. We love the chat between us as we lift and look through the supers.

Sell it all. Sell three hives from the orchard and two hives from the top of the triangle field. Sell the big zinc tank that spins the honey. Sell the smoker with its funny oast-house nose and bellows that can puff a slow burn of sacking. Use the white cotton boiler suits as yard wear. Our bee life is incompatible with our family life.

On the perfect mornings of early summer, when our girls long to be outside, we tell them to stay indoors out of harm's way, in case the bees are disturbed by our handling them. It's not fair. Those girls need their outside days. But I am keeping the leather gauntlets hanging on a peg in the washhouse, just in case . . .

Hoof sticking out, a birthing problem

The morning inspection is not satisfactory, we have a goat in labour, surely, and we all have to go off to school. I tell myself that all will be well, and both girls start up an interrogation about what could go wrong, hoping that I will say 'Nothing'. Michael sniffs in a way implying that it really is our duty to be at home safeguarding our responsibilities, and 'our' really means 'my' duty. But animals give birth unattended. It's a fact.

All day I have a restlessness as if I have correctly predicted trouble, and when we three females sprint into the yard at the end of the school day, we find trouble.

Gloria, a young goat new to motherhood, is lying on her side and we coax her up to see that she is stuck in her labour, with one skinny little hoof protruding from the birth canal. Intervention is required and we can do this.

We parcel out different requirements: hot water with a splash of disinfectant; rubber gloves; clean, old dry towelling; a bucket with hot water and molasses; and a quick consultation with *Goat Husbandry*. The four-page illustrated guide tells me either to push back the hoof in the hope that now the two hooves will come out together, or I have to unhook the bent one with a clean finger. Fully prepared and fully equipped, we march out to become midwives.

We move the other goats out of the way and shut the barn door. Morwenna gives the miserable moaning mother a reassuring pat, and tells her it will be fine. Gloria isn't in the mood to drink the molasses water, but later she will.

I take hold of the little stick-like hoof and push it steadily back where it came from and am surprised to find this is possible. The girls crowd in. We wait for a little while, wondering if the contractions have given up, and then massage her tummy in encouragement. We don't have to wait long before we feel the muscular spasms of her side and her grunts of effort and out slithers both hooves and a small, damp baby goat.

Gloria turns quickly to greet it, nuzzling and snickering and licking, and in a few moments it bleats a response. We help by rubbing it dry and sit back to enjoy the little scene and then witness the effortless arrival of a twin.

A hen at lunch

We have friends coming for Sunday lunch, roast goat leg and assorted vegetables and a rich winey gravy, and then rhubarb crumble and cream; absolutely all of this is organic and home-grown, except the wine and the flour in the gravy. I have half-starved the girls this morning to encourage appetite.

Michael has changed out of his favourite nightmarish sweater into a clean shirt and jeans and Morwenna is into her third costume change. I have given up even looking at what Georgie is wearing currently. I am busy cooking and squeezing everything onto and into my tiny cooker and keeping twelve plates warm in a sink of hot water.

Our friends arrive and we have drinks and chat in the conservatory, kitchen and on the lawn, and the sun is shining. The flower border is ablaze with purple cornflowers, white wisteria, and so many different shades of blue and purple; I have a quiet moment of pleasure gazing at it and floating a little apart from house, family and guests, knowing all is ready and all is well.

With everyone seated, Michael carves and I pass around plates, when I see from the corner of my eye some hens heading towards my blue garden paradise. I know their tricks, the well-dug soil here has a rich supply of worms and they have come to scratch. They know it's not allowed, as I shoo them away every time, with howls and arm-flapping. I have discovered that firing the shotgun out of the window gives a good, scary blast and can keep them at bay for a day or two. Hens can be very forgetful.

I go to the cupboard, surreptitiously open the window and bring our gun casually to my hip and squeeze the trigger. BAAANG. There is instant outcry from the hens, who flutter away; there is outcry too from my surprised guests, some of whom leap up and come to the window, where they can see what I had not planned: on the lawn lies a dead chicken. My guests look at me horrified. Our lunch resumes with a muted atmosphere.

Phee and the teenage party

Have we really done this mad thing? Promised Phee she could hold a teenage party at our farm? This must be because we do not yet have teenagers ourselves and we don't understand the folly of our goodwill.

We have set down absolute criteria, from which Phee has promised she will not waver. She has to know where everyone is, given that they are likely to scatter far and wide. There is to be no drug-taking or alcohol on the premises. There is a definite end to it all at midnight.

Loud music blasts across the valley and it seems that the entire youth of the town has arrived at our place by tractor, on foot, or been dropped off by astonished parents. How weird is it that one of the teachers is giving a party for most of the school?

Our girls are hopping with excitement and rushing about greeting those they know. They have dressed themselves up in suitably groovy party outfits. There is dancing and lots of clustering and chatting and, well, happiness. Everyone is having a fine old time. Michael and I do some anxious prowling for a while and find that our interference is not necessary. We haul someone out of a hedge. Frankie Goes to Hollywood is playing over and over. I see torchlights flashing in the field below, looking like so many fireflies. There are squeaks and shouts and always laughing.

We decide to test Phee out when we next see her and ask if she knows where her friends X or Z are and, wow, she does know, and tells us calmly that X is snogging on one of the hay bales in the barn and Z is dancing in the garden. She sails off again, arm trailed around her own boyfriend.

Just before midnight the parents begin to arrive, looking fearful, and Phee turns on her executive skills, sending various friends to bring back the respective teenagers. They slouch along within minutes and first one and then the next car drives off. We are emptied of partygoers within the half-hour. Phee is delighted: it has been a fantastic success.

As we stand in the dark with a glass of wine, we hear yowls and song twisting up the hill. Our reluctant girls have been sent to bed, no doubt planning their own parties now they have a blueprint.

Studying in the barn

In a few weeks I will be sitting my finals. I have finished my dissertation and now need to revise. The family have been amazing. The girls have painted medieval manuscripts that adorn my finished dissertation and have been so patient and understanding of my time pressures, deadlines and now my desperate need to revise.

So I have come away from life and bustle to find quiet in the vast space of the barn, where I have set up a table amongst the dust motes and a few straggling butterflies. There is a vague sound of poultry in the yard. The rumble of tractors. Distant traffic. An hourly branch-line train. My table and chair, a bottle of water.

No extra anything, apart from today a leather-bound *Complete Works of Shakespeare* and a pack of index cards onto which I am making key notes on the themes and characters and essentials of *Measure for Measure*, with suitable quotations. It's cold enough for a sweater with the sharp breeze from ventilation slits in the stone walls, but the light is good enough. I have given myself a two-hour shift before I have promised to emerge and play, and it's going well. Pressure does that.

I have determined to be a part-time student and a full-time mother, but there have been moments of squeeze in every direction. If Michael takes Georgie to school and I take Morwenna, then I can study until they are ready for collection – sort of. I have only once forgotten Georgie and left her standing outside the school gate with her violin and have paid for that over and over in guilt and apologies.

It is nearly done. Just a few more weeks and then this gargantuan enterprise will be over. Giving myself permission to do this has been one of the greatest joys in my life.

Cows on the lawn

I am waking from a deep sleep with a dream of cows snuffling and breathily heavily and realise it IS cows and they are just below our bedroom window. It's early but not too early, so I rouse the girls and Michael with the alarm call of adventure, excitement, and urgency, touched with neighbourly irritation. Adrenalin works quickly and we are tumbling over each other to get boots on and Michael gives us different directions: two around the front of the house, one through the back-yard gate, and a fourth along the drive and through the top gate. Plan A is to get them off the lawn, plan B to then herd them back up the drive to Jack's farm.

Plan A is already in tatters, as the overexcited bullocks see our team of two around the front of the house and in a mad response, they gallop in a circle around the lawn churning the soft grass as they go, and diving through the clump of fruit trees. There is no way out for them there and as they see the blocked route of the wall, the impenetrable thicket, and the assembly of all four of us regrouped and standing still and formidable in a line, they leap and buck with front hooves dug in and back legs kicking high. We stand firm.

Michael calls out instructions. 'Funnel them out of the back lawn, onto the front lawn and close in behind them. Stop them mucking up the lawn again.' I have a pounding heart, but we all steadily make that line and get them out through the gap and into our field at the side. Morwenna and I then rush down to the side of this field to look ferocious. Michael and Georgie cover the back route.

Slowly, arms spread wide in a preventative net, we force the bullocks up to the road. Michael tells Georgie to nip round to our back gate to block off our own driveway. The bullocks have stopped careering about and are beginning to look confused. Calm and slowly, oh so slowly, we close in. It takes only a few seconds and they are on their way back up the hill, snorting with nostrils wide and flecked with foam.

We keep in steady pursuit, hoping they don't see the gap in the hedge. Steady. Slow. Morwenna and Georgie and Michael keep on up, and I get away to phone Jack to warn him of his little band of returning critters.

Later we wander the garden looking at the churned-up mess. Jump on the ridges? Turn over the divots and stamp them back into the ground? Michael has a better idea, we can form a team to pull the old granite roller back and forth. That could be a long day. Some of us slip away.

Pram wheels

Pram wheels have a long life. Once so diligent in their duty, bumping along the farm paths and up the lanes with the carrycot and a not-sleeping infant, now they are the most dangerous thrill we have to offer.

Two sets of wheels, a fat cushion wedged on the chassis, no accessible brake, spokes whirring – ready to dislodge or dislocate fingers, and a clear run from the gravel and concrete surround of the house and down the steep slope onto the lawn. It's an act of faith and an adrenalin-rush as you hurtle downwards and, at great speed, spin forward along the flat lawn towards the ha-ha and the mess of nettles five feet below.

It's basic physics. The older and heavier you are, the faster, the further, the more likely you are to sustain career-limiting and life-risking injuries. The kids screech and howl as they rush down, legs in the air, fingers always too close to the spokes, faces lit up by the flight.

We line up for our turn. We cajole anxious newcomers to this wild ride. We cram bodies that are too large onto the cushion seat, advise legs held up, push them off and hear the squeals.

Fact: no one has yet gone over the ha-ha. Fact: no fingers have yet been ripped off.

Opinion: best ride around, next to the mud slide. Opinion: not sure how much longer those pram wheels and their brave little suspension system are going to last.

Water hose

Michael has fixed a new outside tap, secured to a wooden post. Its black alkethene pipe runs down into the earth, where it is linked to a new and complicated route to the spring. I am strimming close by. Michael looks dubiously at me and tells me to mind the new tap and pipe.

I am busy with nettles and woody stems and have to swing the strimmer with some force to cut them. I swing right into the alkethene pipe, cut it clean through, and it haemorrhages water in a high-pressured arc over me, the strimmer and the driveway. If I knew where the new stopcock was, I could easily turn it off, but I can't remember and I have to get Michael.

I run to where he is weeding in the vegetable garden. I think quickly about this. I say, breathless and desperate, 'It's okay, it's okay, I'm alright, not hurt, not cut. There isn't any blood anywhere. I'm not hurt, really.'

'What have you done?' asks Michael and I have to tell him.

Chapter 13

Rescuing lost guests

We have finished all the late summer afternoon jobs we could think of and are drifting around, waiting. Waiting for the guests for Owl Cottage to arrive, and then the evening is ours. The girls are roaming somewhere, I can hear chat up in the top orchard. Michael has turned his attention to the onion bed, and I am dawdling and gathering cow parsley and clover for a vase. Actually, I am on alert. I scan each car that I hear descending the hill and see if it wavers and turns towards us. Several waverers have driven on up the valley and I wonder if any could be our guests, who don't properly trust my directions.

When the flowers are finished and on the kitchen table, I start wandering again. The booking form says 'Arrival time 6 p.m.', but who knows what traffic hold-up there might have been. I shouldn't worry really, but it would be nice to get supper on the table. Two weeks ago, it was dark before the Dutch woman and her teenage daughter arrived in a distraught, exhausted mess, having mistakenly thought that a bike ride from Exeter airport to the top part of Cornwall would be easy, flat and fun. The daughter then refused to speak to her mother for the next two days. So I am vaguely anxious now.

I track another car, hurtling along the road too confidently to be a foreign guest looking for a place for the night. Another hesitates, turns right into our lane and then reverses out and carries along the valley road. I feel sure this is them. I clock enough time for them to reach a stopping place up the next hill to recalibrate and I walk into the conservatory and pick up the phone as it rings. 'Have we come too far?' they ask. 'Yes,' I reply and tell them to wait at the phone box and I will drive up and escort them home.

A cluster of eager little faces are peering out of the back window of their car as I roar up, and swing round to gesture that they should follow me.

My worry is forgotten as we all spill out. The girls have rushed down to join in a welcome and immediately take the children off into the farmyard. The parents give that arrival sigh of contentment.

Cottage cooking

I have had a stupid idea and for months we will have to endure the consequences. I have offered 'Evening Meal on Arrival' on the cottage booking form. Of course, everyone driving a squillion miles to some remote part of Cornwall without a nearby Waitrose is going to want it. Want what? Delicious farmhouse stew or lasagne? Which will be easy to have bubbling in a slow oven? Dream on.

It's 7 p.m. and the next guests in Owl Cottage are late. Where are they? If there was bad traffic, they will be grumpy. I have their supper in the oven and it is in danger of being overcooked. It's a roast chicken, roast potatoes, and all the trimmings. After lots of rejected lasagne, it turns out that the only really acceptable food for everyone is a roast chicken and getting the timing right for that is hell.

The guests arrive and are indeed very grumpy, despite my welcome chat. The traffic was bad, they lost my map, they couldn't find a phone box, our neighbours had to flash a torch across the valley to show the lane. I tell them their dinner will be with them in five minutes. Morwenna and I load up the shrivelled roast chicken and the potatoes and the vegetables and gravy and a pudding and it fills two trays. Now it has started raining, so she holds an umbrella and I take the first load, knock on the door, writhe about on the doorstep until they let me in, put it on the table and then dart back for the second one.

What utter madness. It's only August, we still have two more months of bookings to go. Next year the welcome pack will be a bunch of flowers, a cake and a smile.

Denis and the plane

It's a summer day for sprawling and lolling on the lawn with the paddling pool out. It could have been a rare beach day, but there are too many of us here to bother with climbing into cars. We are happy with the smell of drying grass and the hum of bees and distant traffic and the children scuttling around. The pool is filling with shards of cut grass and emptying of water as the children slide in and out. Clarissa and I are lying on an old rug and are propped up on our elbows watching the frolicking children. It's pure summer laziness.

Michael is not with us. I am half aware of waiting for him and the minute I hear him, I leap to my feet and call everyone. 'He's here, it's him, I'm sure.' We stand gazing up at the sky and see the small plane coming closer. As it nears the house, it starts swooping down and then up, up, up, and then nosedives down and picks up again.

We are absolutely silent watching this, not sure of terror or envy. It soars up and the sputtering engine sound stops and for a moment it seems that the plane is in free fall and it turns over and over before firing up again and swooping up. Denis is taking Michael for a ride in his tiny plane and it's a circus show. We have no idea that Michael is sick with fear and holding on for dear life; that very life in Denis's virtuoso hands.

We wave and wave and run about in answer to the skyborne antics and then he is gone, the thin sound of the little plane's engine fades, and we fall in a heap on the spread rug and tell each other what we have seen.

'Did the plane fall? Who was driving? Did you see it turn over and over? How many times?' It fills our afternoon.

Summer night

It has been the most perfect summer day and it is the most perfect summer evening and it is all being wasted in dreary chores and sensible activities and I am grieving. A day and night like this need celebration or at least a special marking.

The setting sun soaks the meadow grass in gold light. The animals are not sure it is bedtime, sharing my reluctance to let go of the day. The kitchen is too stuffy and I need to be out in a rising moon and this exquisite evening. Michael is sitting at his desk marking schoolbooks, saying he is too busy to play.

I take myself off, not into the woods, but up through the triangle orchard and out onto the long field, and I sit amongst the shadows that have lengthened into thin strips down the hill. I watch rabbits grazing under the hedge. Jack's herd of cows are far below, necks bent low, eating. Reluctantly, I gather myself up and return home. Michael announces it is bedtime and there I am, lying in bed and still yearning for the day.

I tell Michael we should be outside revelling, but he is already asleep. I think about this some more and wake him up, and say that we really, really need to be outside. He lifts himself up on an elbow, looking at me blearily, 'What?'

'We have to sleep outside,' I say, 'it's the perfect night.' He points out that he is already asleep. I get a bit tragic and explain that the day is slipping away unnoticed and I long to lie in the field and look at the night. He stares hard at me, trying to work out how serious this is. 'Okay,' he says.

I hustle together shoes, a jumper, a blanket and drag him out before he can change his mind. The geese hear us walking through the yard and start up a noise, which gets ducks and goats shouting too, but we cross over to the gate and climb over into the long field, up a little, and I spread out the blanket on a flattish ridge and we lie down. The sky is now turning from peacock to black and there are stars. I snuggle into the crook of Michael's arm.

Glut of sweetcorn

We have a glut of sweetcorn. It's at its very best, just beyond soft and sappy, but well before dulled and chewy and dry. We are eating them every day, at least Michael and Morwenna and I are eating them every day, with enthusiasm. Georgie is giving me small looks of hate and tells me she can't eat them, not at all.

I load them into the freezer, I give them to friends, but really, we have an embarrassment of sweetcorn. Can I bear to give them to the goats or pigs, or let them grow on and on and then dry them as winter fodder? I ask the greengrocer, whose display is showing some equally fine sweetcorn, if he needs a new supplier, but no. I shall take a large box and put them at the school gates tomorrow.

It's that yellow I like so much. I like ripping down the husks and the silk and finding the rows of yellow pearls. I think mushed sweetcorn fritters, slivers of yellow beads in a stir fry, polenta, yellow cornbreads, chowders, roast hunks of sweetcorn on a barbecue, yellow pickle, maybe dried husks to ward away evil spirits.

Michael and Morwenna and I have sweetcorn for supper, lunch . . . breakfast.

Lost Easter egg

I'm bent double in my flower bed, yanking up rogue suckers of white wisteria and creeping bits of couch grass. There is no point standing up, I might as well remain like a flamingo for several hours, as this job seems as if it will have no end.

Interesting, though, to see the world from upside down. The blue sky inverted and the ground soaring above, and I peer at the underside of leaves in their shadowy state. The wall mosses and lichen are thicker this low down, and then my eye is caught by something glittery in the wall.

It's a chocolate Easter egg wrapped in pink foil, a survivor from the garden Easter egg hunt months ago. What a trophy! For full appreciation of this, I stand up and stretch my back and give the egg a close inspection. It is not seeping, or nibbled by ant or mouse, and the foil is nearly intact. I unwrap it and put it in my mouth and savour the extreme chocolatiness of the situation.

Eclipse of the moon

Twilight has slipped away and we have the deepest of blue, early night skies, and an excited bustle of activity on the lawn. We have assembled a small party to watch the eclipse of the moon and we are bringing out chairs and rugs to the part of the lawn that will offer the widest viewing range of the sky.

I have made squares of pizza. Liz arrives with her brother Bob; Michael is telling us the exact timings of the stages of the eclipse, but I am caught up in the thrill of the moment and want to run in wild circles and sing strange songs. This I must not do. It is a serious scientific moment of awe and wonder and any foolishness from me will spoil it. Morwenna is busy asking the right questions.

The moon has risen large over the wood opposite and it's odd to think that soon it will be covered by a greater being. I let my energy run loose by rushing in and out of the house to bring useful things to the gathered group: binoculars, my camera, wine and glasses. The atmosphere is very good, everyone upbeat and eager, being scientific, chatting happily and looking up every so often to check the unconcerned, unaware moon hanging silver and remote above us.

Then it happens. The slightest of shadows has edged away a part of the moon in a grey fuzzy line. It's exciting. We stand or lie back in chairs with our heads craned upwards, passing binoculars, watching. The moon shrinks, dwindles, disappears before our very eyes and we become silent, and then before the primitive mystery of its disappearance can grieve us, it fights back and climbs out of the shadow to hold its place in the sky.

We are cold and go back into the house, where Bob swivels onto the piano chair and plays the *Moonlight Sonata.*

Missing pigs

This time it is Georgie's spoiled and tickled pigs, Gwendoline and Cecily, who have escaped.

Louise has a Vietnamese pot-bellied pig called Vera and she had noticed strange behaviour, seeing Vera rush across the yard into a disused shed with mouthfuls of clematis that she had torn off the walls. Too busy to investigate, she then noticed Vera's absence from the back door, where she usually insisted on tripping people up, hoping for food. Vera was deeply busy in childbirth and had made a bed out of the clematis and was found suckling four hybrid piglets, half-Vietnamese and half-ginger.

Louise was puzzled, and remembered that her neighbours had briefly had a pair of ginger Tamworth pigs to fatten and guessed that the enterprising Vera had found a way to intimacy through a chain-link fence.

Georgie and I were in raptures at the beauty of ginger-black pigs and willingly gave them a home when they were ready to leave their proud mother. They had silky black hair and freckly ginger skin and instead of the pug-nosed face of their mother, they had elegant slim noses.

But they are escape artists. We found ways of sealing down the edges of the link fence, blocking escapes, and persuading them that life in the yard with good food and deep straw could be enough.

But today they are missing again and the yard gate into the field is open. We let out all the other creatures in the normal way and set about looking for our escapists in the big field. Our animals – goats, cow, ducks, geese and hens – are familiar with each other and know the hierarchy and territory allocated to each. They mostly ignore each other, but today there is no need to search for our missing pigs, as we are being told where they are by the creatures of our farmyard.

Standing at the gate, we can see what looks like the widespread field of a rounders game. There is Rosetta down by the river with her head lowered and pointing mid-field, two goats have taken up positions making a semicircle, and a line of geese with their long necks outstretched menacingly are pointing at the two pigs, who are pleased with themselves and busy rooting in the earth for worms.

Georgie, Michael and I spread out to fill the gaps in the circle and start to close in on the pigs, and amazingly so do Rosetta and the goats. We all move towards the pigs, forcing a clear route back to the gate. Rosetta keeps her head down in a ferocious way, as do the goats and geese. No pigs are going to get away with trespass in this field. Not until the gang is quite close do those greedy pigs notice that they are being hustled back into the yard and

Cicely gives a little squeal of shock. We close in and shut the yard gate and shoo them into their pen.

We go back to see what the animals are doing now, but – game over – they are all grazing as if nothing had happened.

Gleaning

After the hay harvest, the world feels full, comfortable. To walk past the open barn, where the bales lie stacked as neat as bricks and steaming gently, makes us feel as rich as Solomon. We did it. We waited and clung onto the hope it could be done before rain, and we did it.

Now in the early morning we look out onto the field, which is strangely denuded and pale and stubbly. The morning mist is rolling seawards down the river edge and Jack's cows grazing in the next field have lost their ankles in whiteness. We wander towards the gate to lean and gaze, and see that this simple journey kicks up wisps of hay not picked up by the machines. We have some gleaning to do.

When the sun has dried off the overnight mist, we work as a team to catch the loose wisps of hay in sweeping motion with rake and forks and the hayfork. We are a pastoral oil painting: a strong man in loose trousers and a head of black, curly hair; his wife in dungarees and a flowered blouse; two girls, blonde and curly in black boots and blue boiler suits; a scattering of ducks, geese, goats, and butterflies; a wooded copse, a reedbed, the river. In the distance stand the high arches of the viaduct and a line of poplars. The sky is blue.

This must have been painted over and over and over; we are a timeless rural scene.

The unending drive project

Michael has arranged for a lorry to bring vast mountains of gravel. 'No, not gravel,' he says, 'it is 3/8ths to dust.' A lorry arrives before I have had even a morning cup of tea.

We hurry out to manage the lorryload of stuff and persuade the driver to turn and reverse down the drive, depositing gravel at intervals along the length of it. How Michael and the driver have worked out how much to tip at each interval impresses me, and I see in a flash that this has saved us serious amounts of labour.

We stand about and watch Michael start up the rented steamroller, which is nothing more than a colossal drum that shakes violently and is attached to a handle. It shakes and flings Michael from side to side. This is going to be a hard beast to control. He turns it off, visibly shaken, and commands us up to the far end of the drive to the first pyramid, handing out shovels and rakes.

Two shovel handlers get to thrust the blade into the stones and scatter, fling or throw the gravel onto the muddy drive. The rake handlers grab stones with their rake and pull them onto the drive, and then have the job of smoothing out an even surface. We are all quite jaunty about the job, as it is rather fun and the sense of achievement is fantastic. We have Morwenna and Zafar and even Georgie full on with this task, for the moment. Michael soon abandons his role as shoveller and starts up the violent steamroller, and then he and Mark squash the gravel down firmly into the ground.

Our euphoria is short-lived. We can't chat and lark about with the noise of the steamroller drowning what we say and by the time we have spread out the second pyramid, we have all got backache and sore arms and the beginning of blisters on our hands. I call out for a coffee break and find supplies of gloves and some cake. Eight more pyramids to go.

Rescuing the hay barn

After a stormy night, the hay barn is in trouble. It is a structure of long pine trunks supporting corrugated iron sheets as roofing, with one side made of stone wall that cuts into the sloping top orchard. It's big enough to hold the entire hay crop, a few years' log supplies and once Amelia's car. But last week's big winds have wobbled the end section and two of the supporting poles are teetering. There is a worrying creaking and flapping from some loose corrugated panels.

Chris, the ex-student, igloo builder, and now roofer, comes round and gives the project a hard stare. We love Chris and his extreme life and spend hours across the kitchen table with tea, listening to tales of his solo boat trip to Iceland with only a sack of potatoes for victuals; or drifting down a river in Gambia; or lying at night on the Cornish mudbanks, with rats running up the mooring lines and his slashing at them with an axe while half asleep in his sleeping bag. Georgie has decided that Chris is a Real Man.

Today it's roofing business, and he is up on a ladder looking at the strength of assorted poles. We will have to cut the barn in half, he decides, using the good end and the good poles and making an overhang of corrugated iron with some of the panels. And he is up, wielding a hammer and showering corrugated panels to the ground like a guillotine. Women and children are sent inside and Chris and Michael work with loud hammering.

Before the day is out, we are summoned to look at the new, smaller haybarn, transformed: snug, smaller, dry, still big enough for hay and logs, and let's forget about cars.

Ducks needing rounding up

Last night the ducks didn't come home. We went down to the river and searched up and down, but found nothing and left their shed open in case they came in after dark. I think we knew that their return was unlikely.

But this morning there is new vigour and a search party.

The first thing is to have the whole family, like the Railway Children, down by the level-crossing to flag down the little train on its way to Looe. The driver sees us waving urgently and stops the train and pokes his head out of the window. So do all his passengers. We explain that a flock of ducks, some fifteen or so, has disappeared and can he look and see if any are on the river along the railway down to the sea.

We then take bikes along the rail track, past the old signal box at Coombe, to investigate the watery remains of the old canal. Morwenna hears the quacking first and we push our way through the fence and see our happy escapees. But how to get them back home? Ducks are swifter in the water than four people in wellies.

We cycle back to get the right equipment and return with thigh-high waders for Michael and several long sticks.

Michael fights his way into the canal, and we stand upriver, and with swishing and splashing we drive the flock out of the water and up into the bank. We hurry to rearrange ourselves behind them and persuade them onto the road, walking behind them at a leisurely pace and keeping them on task with the sticks.

We have only half a mile of road to cover and hope they won't be sidetracked by the river hidden by undergrowth, but the ducks seem happy to be driven along and quack cheerfully until a car comes up behind us. I dart to explain to the driver that there is going to be slow progress, our ducks are on the way home after a night out, and the driver wisely reverses and finds another way.

Finally we are at the turning to our land, with a muddy slipway at the bridge down to the river and our own safe meadow. The ducks sway and waddle down the slipway and launch themselves into the river and swim fast downstream.

Sex talk on the lawn

It's time to have the big talk about sex with Georgie. I don't know if she will be wildly embarrassed, or superior and knowing. I am planning this very carefully. The day is a rug-and-picnic sort of day and I ask Michael to keep Morwenna occupied while Georgie and I talk the talk.

I lure her out, oh so casually.

I appear in her room with the rug clutched to my chest. 'Come and laze on the lawn and thread wild strawberries with me, I've found zillions in the bank.' Irresistible.

The berries are in a rhubarb leaf on the kitchen table and we find the right stems of threading grass and three of us sprawl out on the rug. Woo the cat is content to stretch and purr as we thread strawberries.

When we have lined up our little feasts and then sucked the berries off the stalks, I ask Georgie to look at the cartoon book on body changes with me. We turn page after page, discussing it all and I think it is going okay. Some questions, some answers, no embarrassment, some comments that show me it is certainly the right time.

Then it's over and as we return, Morwenna smirks and says, 'Had the Talk, then?'

Washing in the back yard

There is competition for space in the rigging; the back yard with five long lines of intersecting washing lines has become a three-masted galleon, currently swathed with a spinnaker in purple, green and blue that Michael has washed and pegged out, and it is in extreme danger of picking up the house and flying away in these gusts of wind. I push my way through the billowing bulges of fabric to see if there is room for my much more important triple-load of sheets and pillowcases and duvet covers from the holiday cottages; and there isn't.

I clench my teeth. 'Sailorman!' I call out. 'I have all this cottage washing to hang out – can you finish off the spinnaker in the sitting room over the sofa?'

He comes out to the back yard and we stand there, each looking at our respective projects and then at each other. A pause. I can see Michael weighing up my need for the washing to be done and dry against his more pressing mantra of 'The Boat Must Come First'.

He is silent. So am I. Then he gives a little nod and says, 'Help me gather it in then,' and I do, with a kiss.

Puffball

Morwenna has found a giant puffball mushroom growing in the greenhouse, the size of a football and gleaming white. It has a perfect smooth and almost tacky surface, and a small, uneven and earth-pocked area where it has been attached in the ground. It just appeared overnight. It is surprisingly light and when cut, it squeaks like polystyrene.

Morwenna cuts half of it into slices and proudly prepares a pan with butter; we fry it and then are disappointed at the slimy texture and weak flavour. I pack the rest in the fridge and the next morning I fry it in bacon fat with shards of bacon, and this time get it properly browned. I add small tomatoes and it is a feast.

A purple bank

If we stand on the ridge of the sloping field and look across the valley, the bank of woods opposite has changed from wintery brown to purple. We don't understand. The wood is a mix of alders and oaks and is green in summer and brown in winter. That's the rule. So why are we seeing this giant whale of a woodland carrying a flush of purple? We need to investigate.

At the fork in the road, where it turns off over Trussel Bridge, there is a chance to scramble up into the wood and so we do, past the place of badgers or orchids and into the tangle of roots and the fine, rich mustiness of leaf mould. Everything is looking properly brown here. Somewhere there must be purple.

We reach and stretch and find some fresh young fingers of growth and yank them down, peering at them with proper scientific and artistic interest. There is the solution: the alder buds, still tightly sealed in their winter packets, have a purplish tinge. We conclude that one million million little purplish tinges waving into the air must be sending a purple glow to us over the valley.

We are satisfied.

The mud slide

Oh, the mud slide! It's what we do when an adventure is needed, or cross words or surplus energy need to be burnt out. This is how we do it.

Put on wellington boots and stride out down the drive, through the cottage courtyard into the vegetable garden and over the gate into the field. Then the walk becomes one where we must follow the ridges made by cattle or landslip to get a thin, flat path, rather than an ankle-turning slope. Along the middle of the field the land falls down to the river valley, marshy and livid with irises, and it's the middle contour we need for the mud slide. Ten minutes of fast walking takes us there. Some old forestry land straddles the river-bed and only a few metres deep into the woodland lies the wrongly named mud slide – because it is actually a rope swing.

The turn-taking involved is agonising, because everyone wants turn after turn on this ride of sheer adrenalin. You slither down the muddy bank and grab the rope with its small wooden bar, step back as far as possible, and then dig your heels into the mud to prevent slippage. Then twist onto the wooden bar seat and let go. The rope swings out across the river and towards the opposite bank in a wild rush; you are high above the river, which glints below, and with each swing your pendulum decreases.

Will you leap off in time or spill into the river? Land in the mud? Or neatly step off with a well-timed lunge? Then give up the wooden bar to an eager hand behind you.

Today a gang of us arrive to find the rope cut by vandals. We march back to assemble a repair kit: some old climbing rope and a length of baler twine, a metal nut and a catapult. We tie the nut to the baler twine, the baler twine to the climbing rope, place the nut in the catapult and fling it up into the boughs of the oak.

We miss only a few shots and then the nut with its cargo of baler twine whips over the strong-arm bough, we pull down the thicker rope, and reinstate our fairground ride: the mud slide.

Chapter 14

Hedgehogs marauding

Someone is coming in through the catflap at night and upsetting dishes of dried cat stars and leaving poo. I ask the cats about this, and we look together at the mess and they give worried looks with big eyes. The girls and I become detectives and look for further clues. This is quite difficult, as the conservatory and the washhouse are both full of stuff. We agree not to turn everything upside down, but be observant and vigilant.

Georgie finds another poo and we bend down to inspect it. Too small for fox. Definitely not rat, because rat poo is oval like olive pits. We keep looking until Morwenna sees the culprit curled up between the washing machine and the wall, pressed close to some insulated water pipes.

We have a snoozing hedgehog.

Heron rescue

A strange commotion on the railway line. An indistinct sound of flapping. David, our most loyal holiday visitor, comes puffing to the back door and says he has seen a heron on the railway line. Morwenna, David and I hurry down, pushed by the anxiety that a train is always an imminent possibility. At the bridge, we teeter over the slanted sleepers and walk south on the coarse chippings that lie beside and under the rails. When we are almost level with the farmhouse, which lies stark against the backdrop of the hill, we can see movement on the track.

It is a heron lying at an angle and it is flapping one very large wing. Obviously injured, it looks both desperate and terrifying; I had not understood how enormous such a creature could be. We stand futile and wonder what to do. 'We have to get it to the vet,' says David, and while I agree, I cannot quite see how.

We walk back and share ideas as we go. How long do we have before the next train? Is the heron close enough to the track to be further injured or killed? We think not. Together we gather up rope and a quantity of garden netting and I ring the vet to check that they will welcome a large, very cross heron. The train is not due for twenty minutes and I drive us down and park by the bridge.

We approach the casualty with care and perform an impressive net-throwing action to immobilise it and we gather up the panicking bird, all the while keeping out of range of its long, razor-like beak. David gathers up the netted parcel and holds it beak forward and then we fold and tuck our casualty into the boot.

We deliver it to the vet and drive away, not wanting to know what fate awaits the poor heron. As paramedics, we have done our part.

Louse powder for goats

We do not have sheep. We think they are too demanding of care and attention and shearing and dipping and they have maggoty tails. We have goats. Lousy goats.

We have a large tin with a handle and a perforated lid, looking as if it was made in 1950. The plain white label stating LOUSE POWDER is stained and peeling and the tin is almost empty. There is every reason to suspect lice on the goats; they are irritable, flicking tails and twitching and scratching, so with a spotty hanky tied over my nose and mouth (and therefore steamed-up spectacles), I take one goat by the collar and start vigorous sprinkling of powder, all along the back and over the sides if I can reach.

I hurl the tin to one side and then look away to avoid breathing in the noxious cloud of powder as I try to rub it in. If I do it outside, the wind blows it away. Wedging the spluttering goat onto the manger and freeing one hand, I pour stuff in my hand for under the belly and between the legs. The powder is suffocating us both and I am trying hard not to breathe it in. I release the goat and we both dart outside to gasp real air.

Later we will trim toenails. Then there is grooming with a strong brush. And the medicine for mastitis. And providing a salt lick. Cobalt tablets. Searching for lost collars. Taking females to the billy goat.

Our goats are yellow-eyed, charming, funny, wise, wicked, soft, and affectionate, and our girls lead them around, scoop up the babies and cuddle them, bottle-feed them if needed, and treat them to tempting morsels of leaf and branch. We are a family of goat lovers. No sheep.

Shovelling manure

Maybe thousands of hours of my life have been spent loading manure into wheelbarrows. And Michael even more so. Is it a noble occupation? I have my efficient method with a long-handled Cornish shovel and long-handled fork. The unfouled top layer of the old hay becomes the base layer of the future. Then the next layer can get wheelbarrowed away to the heap by the duck shed. When it's high, we run a wide plank up to the top to get an easy run to the summit. We take turns to do shovelling or barrowing.

We are dressed in boiler suits and wellingtons and are red-faced and sweaty. The girls join us or leave us as they like, often working harder than me. Wearing gloves helps, but we will blister anyway. The smell of the manure is not unpleasant and the goat dung comes in small pellets, not wet slop, but it gets compacted over the week.

It's the pig-shed manure and the duck-shed manure that are gross. This job is on offer to anyone who wants it, and Michael's sixth-formers often come to join in a morning of work. They sit in borrowed boiler suits chatting physics and chemistry in coffee breaks, teasing our two girls.

The best moment is the renewal of the hay layer, scratching back into place anything saved that was dry and sweet-smelling, and then bringing in bales from the barn, breaking them up and threshing them about on the floor.

No, the best moment is leaving the yard and peeling off the boiler suits and going upstairs for a shower. Think proud Breugel peasants, if you like, but it's not a noble occupation.

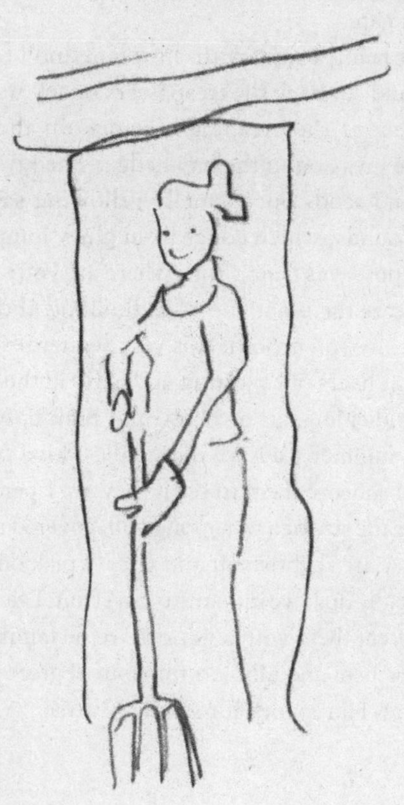

Kiwi fruit

Why did we give the most sheltered growing position on the entire farm to a rampaging inedible fruit? We could have had white peaches with pink downy cheeks hanging low and heavy on their branches. What about apricots? Maybe we were aware of silverleaf and other peach or apricot miseries, but the choice of a flamboyant kiwi fruit now seems inexplicable.

I don't even know if it is a bush or a tree. It has thick climbing stems that reach out year after year and send tightly wound feelers out and round my washing line. It has climbed up the wall, out and over the wall, and is progressing aggressively to fight with the bay hedge and the garden gate.

I am not being a sissy with this plant and I take out my secateurs and cut back the trespassers on my washing line, I hack away at the creeping advance on the gate, and untwirl the invasion in the bay hedge. The kiwi is furious with me and sends out a smelly yellowing sap from the pruning wounds, which congeals in gluey lumps.

The purpose was fruit. Fruit, where are you? Surely after multiple years the plant can start thinking about fruiting, and we resolve to uproot it this very winter if there are no fruits. Ha! It hears our plotting and early in the spring puts out some ridiculous little yellow and pink flowers; by the end of the summer we have the smallest kiwi fruits imaginable and I squeeze them to see if they are ripening.

They are the size of a ping pong ball, covered in green fur, but yes, they are slightly soft and when I pick one and cut it open, we really do have miniature kiwi fruit. I gather a bowlful and present them with a flourish to the family at supper. We quarter them and nibble a tiny bite of greeny flesh from each quarter. Hard work for small rewards.

Smelling the morning

It's my drug. Smelling the morning. I open the conservatory door and step outside and breathe in a huge gulp of air. It is our own narcotic air, fresh, green, heavy with the tang that every different morning brings. It could be hay ripening. It could be dung spread on the near fields, or the waft of daffodils, a new leafy smell or musty rotting leaves, tractor fumes down the lane, anything.

I suck it deep down and hold it in my chest or in my brain and it lasts me all day. And then I drive to work.

Running the bounds

I am very definitely giving up on the gym after two dismal episodes. Firstly, I was sitting on the machine rowing my boat with a fine old speed and rhythm going, when I was spattered with rain, puzzling until I saw that the man on the running machine close by had drops of sweat spinning out from him and I was the unlucky recipient of too many. I left.

Then I got caught out on a fire-alarm episode when Michele and I were having a sauna, and the lifeguard came up to the steamy cabin and told us to leave without delay. When we said we couldn't go outside on a November day in our swimming costumes, he told us there was no time to dress and we would be given covering blankets as we left the building, please assemble in the car park. We were given a small tinfoil blanket each and stood in the car park covered in goosebumps looking like turkeys ready for roasting.

But there is a good alternative for fitness; I see it is entirely possible to have a running circuit along the bounds of our own fields. Some parts are soggy and need a detour at those times, but there is a good section in the bluebell part of the top orchard and a great dip in the small field where Rosetta likes to give birth, and a double loop around the lawn and a stretch below the ha-ha; running along our own paths is amazing.

I can see the fronds of bracken unfurling, dodge past clumps of bluebells, duck under the hazel branches, watch the progress of blossom that will turn into apples and pears, check how high the meadow grasses are growing and fattening, and breathe deep, breathe deep; I snag my legs on bramble serpents lying sneaky in the ground and answer the calls of the goats, sing loudly, bend over to catch my breath and see ants, nip in by the rickety gate, and then run along the new boundary into the new field, down along, along.

Midnight cat

There has been too much schoolwork tonight and my brain is fuzzed. I ask Georgie and Michael if they want to come for a night walk. Michael has too much still to do, but Georgie throws down her revision gladly and we pull on boots and a jacket against the heavy dew and are out of the house fast.

We don't take a torch and the night has a loom about it, which means we can almost see; it's less than twilight, but it will do. We are up and over the barred gate into the wood when a thing streaks past. We have an idea of who that thing might be, and I call out, 'Woo?'

We walk on, breathing in the sweet smell of the damp plant life, knowing our way well on the stony path. We can hear a faint rustling sound. As we reach the top of the slope where the path flattens out, a thing hurtles itself out of the higher bank and lands with a thump in front of us. If it was a small boy, it would be wearing a fierce face of menace; but it is a small boy cat, Ezekiel, who loves a good pounce.

We bend and fuss over him and he twines about our legs and then rushes off to hide.

Meat day

I dread Meat Day. The children are at school, Michael is at school and I have gone to collect one whole butchered bullock. I pretend I don't know his name or remember his soft breath on my cheek. He nearly fills the van. Joints of meat, all neat and rolled and tied up with string, are crowded into some heavy-duty bags; the rest are full of unidentifiable tangles of meat-coloured stuff.

But the butcher's in Tideford is a jolly place and they have happily chatted about what I have in my assorted bags. Some of the cuts I will have to go home and consult Mrs Beeton about: boxeater and some other things unknown to me, but helpfully labelled.

I pay for the butchering and drive home and then, as I start unloading, the job suddenly feels overwhelming. The big bags are heavy, and I lug and bump them down the granite steps and into the kitchen and leave them on the floor, one by one, propping each other up like drunks. Cats Woo and E come in and start shouting enthusiastically.

I unpack the important joints first and stack up sirloin and topside and more sirloin, and more and more and then briskets, until the table is groaning. It looks seriously more than we can manage, even with a roast beef Sunday lunch every week of the year. I make a few phone calls.

Thirty labelled joints of various kinds go into the freezer, followed by neat pairs of steaks. And it's already school run time. I shoo out the cats and whizz off.

I spread the word amongst the school run mums that I have beef to spare.

When I get back, my kitchen is swarming with women, eager for organic, home-reared beef to feed their family. The girls join in sorting through what people ask for and I conjure up prices. I haven't bought meat in ages. Carrier bags full of meat are walking out of the kitchen, money is changing hands, and the table is clearing a little.

When the crazed rush has subsided, we survey the scene and see that there are still two silent, lurking plastic sacks. We open one and peer in. Huge plastic sacks, entirely full of mince. It is not in bags. The meaty bloody smell is overpowering, and the most enormous mass of meaty mince is waiting to be . . . what?

We set up a mince processing plant, the girls and me. Georgie is in charge of opening bags, I am in charge of plunging a hand deep into the mince and bringing up two 'gurt big dollops' for each bag, and Morwenna is in charge of placing the cleanish edges of the bag into the heat sealer and pressing the knob, which makes a most satisfying purr.

After the gruesome business of processing one sack, my team have had enough. Later that night, I resume the meat processing and Michael joins me. I plunge, he seals. I will do more tomorrow.

Bath time was never more needy or more fragrant. I lob bath essence in extravagant quantities into the water and scrub and scrub like Lady Macbeth. My dreams are pungent with the smell of blood.

Papier-mâché

I brought home a trophy today, from the art department at school. They were being thrown out and I rescued them from the rubbish heap; two life-size, papier-mâché figures made of chicken wire, but entirely covered with robust paper and then painted appropriately as a schoolgirl and a bus driver.

They upset various car drivers on the way home, who did not understand these dangerously entwined figures lurching across the back seat, and I daresay they are now about to upset the postman, the meter reader, the Jehovah's Witnesses, and any other stranger to the house.

I have positioned them in dialogue, the schoolgirl at the conservatory table seemingly pointing at something and the bus driver on the wooden box by the washhouse. They are perfectly in harmony. I bring Hubert, the stuffed dog on wheels, out to join them and he stands on his little rubber wheels close to the bus driver, a loyal hound with the most tolerant and benign expression.

As I stand back to enjoy their silent dialogue, I have no idea of the fancy lives they are about to assume. I shall witness each of them being spoken to by holiday visitors. They will don medieval costume in a church pageant, they will roam the house and assume unexpected positions. And in a final gesture of self-sacrifice, one now tattered and torn will became Guy Fawkes, and we will watch tenderly as the bus driver goes up in flames.

The rat in the ivy

I put it there myself, a witticism, wondering what real use a life-size, black plastic rat could possibly have – once the thrill of pulling it out of a Christmas stocking had worn off. It was tucked into the variegated ivy that has taken hold of one side of the conservatory. At the time, it was quite well tucked in, with only a bit of tail and a vile, ratty face showing, but over a season of ivy growth it had almost vanished.

I was busy with a tidying project in the conservatory and had newspaper on the granite slab floor, a pot of duck-egg coloured paint waiting, and I was wrenching off strands of ivy to detach the little suckers by which it clung. I was stretching around the geraniums, which are currently man height, so nourished are they by the rotted compost from the midden, and trying not to break off any branches with my sharp jerks, when out flew the plastic rat past my face and landed with a lifelike thud on the floor.

Well, it fooled me properly. I think my heart stopped beating with the horror of this huge rat in my home and I lurched back to get my balance, flailing around to avoid stepping on or near it. It remained motionless on the newspaper and the world returned to normal.

Now what is one supposed to do with a plastic rat? Is it okay to lob it in the bin, because that is exactly what I did.

Things in the freezer

Every time I open the freezer there are two bags of things I can't abide sliding into or out of my grasp. I need to get rid of them. One is a cluster of four trotters from the pig we slaughtered. They are like the lumpy handles of a skipping rope and I just know I am not going to find them the delicacy I should. And worse is the large bag, mercifully made indistinct by ice crystals, with the grinning half head of the same pig.

Michael and I slip out of the house after dark, feeling like grave robbers, shouldering a pick axe and a long handled shovel and with a torch. It's just as well we have no neighbours or the police would surely be called out. We are holding a sack containing unwanted body parts. We walk deep into the woods and dig a hole, cutting through fern and roots and shovelling away earth and stones. We work steadily and quietly and empty the sack and arrange head and trotters and cover them up and tread the earth down. We will have to return later to check that the fox has not dug it all up, ready to horrify all forest walkers. We are a gastronomic disgrace, food cowards.

Goose crime

A phone call from a neighbour. There is a man with a gun who has just shot one of our geese and is walking towards Trussel Bridge with it under his arm.

I shout out to Michael and the girls, 'SOMEONE IS SHOOTING OUR GEESE,' and I'm out running across the lawn, down over the ha-ha into the field in hot pursuit.

The poacher is far off, but I can see him and he is walking steadily, but I am running. I reckon I can catch up with him; I am fuelled by rage. I have to do this with stealth, so I keep to the riverbank, where there are enough alder trees to give me some cover. He hasn't noticed me and on my left side in the river I can see two geese in a panic swimming back and forth. I wonder whose wife or husband this poacher has taken.

I'm getting closer, but I stumble in the uneven patch of wild irises and twist my ankle. I keep going and the poacher is in sight with his humped white burden and his gun. How dare he? And what for? Maybe he has a starving family to feed, but that goose is part of our family, part of its own family. He still doesn't seem to be aware of me creeping up on him, but neither does he seem worried that he might be seen, he is simply walking. He is perilously close to the woods at the end of the long field and I could lose him if he gets there.

There is loud shouting behind me and I look back to see two policemen running, they are coming my way, but they are shouting at me to stay where I am. They huff and pant up to me, 'Stop, madam!' I am unarmed; they say, he has a gun; I am in a confrontational rage and who knows what the outcome of our meeting might be. I'm aghast. I had not for one minute considered that I might be in danger.

I am escorted home by the policemen, and I am properly told off. I learn that Michael rang the police and they were outraged that I should have pursued an armed man. He was chastised for letting his wife behave in such a way. They do not seem in the least interested in capturing a poacher, who in past times would be hanged or sent to Australia for such a crime. Apparently, I need to review my impulsive nature.

Parcels

Brown paper and string. Parcels by post, rows of stamps, maybe even sealing wax. We seem to be in receipt of a succession of charity parcels from Michael's mother, who sends salami and frankfurters thinking I do not feed him properly, and from his sister, who sends exotic clothes she cannot pass on to designer second-hand stores.

The salami is sometimes started first by the cats, who break through the brown paper and munch if the post arrives while we are out. The mystery sister parcels are Christmas and carnival rolled together. Clothes quite unwearable in our Cornish market town are pulled on and swished about by us queens and princesses of dressing-up. Leather skirt with huge buttons and exaggerated kick pleats; gaudy silk chiffon, floating negligee; shocking pink bomber jacket; the range is exotic and scented with money.

Today the parcel is smaller but heavy. I wait till the girls are home from school before we break open the box and out spills a pirate's treasure of jewellery, bling, fat fake pearls, Moroccan beads, enormous rings, a thousand bangles, and turquoise plastic jewels.

The day is grey, but inside our kitchen we are gilded lilies.

Putting on wellies

It is never wise to put on wellies in a hurry. Not ever, ever in this world where wellies are kept either in the musty, spidery, cobwebby washhouse, or more frequently in the conservatory. Here a myriad of cultures and events are at work. The aphids on the abutilons are giving off dewy droplets much loved by ants or mould; the vine in flamboyant blossom is eager to drop petals or, when greening and plumping its grapes, will let fall the failures; and spiders who live in the vine will wrap leaves that drop. In the autumn, leaves crinkle and fall of their own weary accord, and by winter, the few unpicked and rotting grapes heavy with purple sweetness will crash in clusters.

All this dropping, crashing, floating, falling, and creeping can be certain to take place in the gaping mouths of the wellington boots that stand in untidy groups by the back door. And if you are so foolish as to let your boot lie slovenly on the floor, then it may be hosting a dead or dying mouse brought in and tortured by the cats.

So sit slowly on the sofa and turn your boot, and shake out the hard, green grapes, the mouse, the crackly leaf, the webby bark or the purple rot. Your foot will not know what horror it might have squelched. Take your time.

Septic tank

Lurking at the back of my mind is the uncertainty about our waste system. Somewhere out in our acres is a septic tank, a soakaway, a something which never showed up on the map with its elaborate twiddles and fancy seals from the Duchy of Cornwall when we bought the place. I even approached Jack on the farm above and asked an English question: Do you know where the septic tank is? And got a Cornish answer: No.

On this day there is major activity in the field. We have given permission for a new waste pipe to be laid from St Keyne and diggers are busy with a huge trench. One of the guys is labouring up to the house looking puzzled and he reports that they have broken a clay pipe running down to the river. I'm delighted, it might solve the mystery: will this be the runoff pipe from the septic tank?

It requires a logical approach. I am to stand by the broken clay pipe, Michael will flush a loo and yell at me at the same time and Georgie and I have a stopwatch to see how long this flush takes to reach the pipe. Morwenna pours a whole bottle of blue food dye and black ink down the loo, and then there is The Flush, a big shout and some waving out of the small bedroom window.

It only takes a few seconds for a torrent of black water to reach us and with it the terrible truth that for fifteen years all our household waste has been rushing directly into the Looe river. I report back to Michael and we stare gloomily at each other; there is a vast financial implication and a horrible ecological history here.

I ask Michael the Cornishman why Jack said he didn't know where the septic tank was and Michael the Cornishman explains to me. Wrong question.

I should have asked, Is there a septic tank? He would have answered No.

Decorating

I wander with a small brush, dabbing at marks with a glop of whitewash. I cover up splashes of cooking mess with a paintbrush, I erase children's drawings with a paintbrush, and I tease dusty corners clean with a paintbrush. It's modest, because I have learnt my lesson.

I cringe to think of that great plan I had to repaper Morwenna's bedroom, where I stabbed a scraping tool under a lift of paper and brought down, with a dramatic sweep, several stiffened metres of swirled seventies paper and significant amounts of plaster. The wall behind was battens and crumbly lime mix and I had unleashed a disaster. Lumps fell on the floor beside me. Clouds of masonry dust flew up.

Horrified, I rushed to reassemble the wall with filler and used my hands to pat in the goo, smooth it over, and pause to wonder what to do next. The new Laura Ashley wallpaper clung to the lumpy contours in bulges.

Then there was the spill of gloss paint on the stair carpet, not drips but the splurge and glug of a whole tin, seeping horribly and forever into the weave. Replacement required.

So when Rae said she needed an old-fashioned bedroom for a stencilling project and could she use ours, with its Victorian bed and its embroidered white quilt, for her to stencil one whole wall with light blue and turquoise flowers and then photograph the whole as advertising for her stencil business?

Well, yes, yes, yes, yes, yes!

Plucking Christmas geese

Clarissa and I have opted to do this together and it is a heroic undertaking. Two geese are required for Christmas dinners: one for us and the second sold to a neighbour. We have borrowed a Burco boiler and set ourselves up with two stools, aprons, rubber gloves, and black plastic sacks. The children are watching television, we hope, rather than helping Michael kill the geese. Actually, his is the heroic undertaking, as the geese are his friends and trust him.

We busy ourselves with kettles of water to go into the boiler, as I can't find an extension lead long enough to go into the back yard, and Michael arrives with a headless goose as the boiler is half full. The theory is that hot water on the skin of poultry will mean the feathers just drop off.

Needless to say, the theory is wrong or we aren't doing it right. The first problem is that goose feathers seem to be waterproof and when I push the goose into the boiler, it tries to bob up again and no water penetrates its shiny white coat. But by plunging it vigorously up and down, some water gets into its fluffy down layer and a lot onto me. Clarissa runs in to get another kettle-full and I keep plunging. We manage to scald parts of the bird with boiling water poured from the kettle, while I hold it up by its pink webby feet. The warm water that falls onto my wellies feels delicious.

We decide to start plucking.

We sit opposite each other, knees touching with the goose on our laps, and start wrenching out feathers. They absolutely do not fall out. The big pinion feathers are set in concrete. We mutter darkly and roll the creature over and have a try on the breast feathers, which are smaller. This is better and we pull off small clusters of feather and fluff, but our fingers are wet and the feathers stick. The idea of the black rubbish sack is hopeless: the feathers are either glued to us, or if shaken are picked up by the wind and deposited anywhere.

We are both grim-faced and clenched and start grabbing and pulling feathers in a brutal madness. Feathers are flying everywhere. The wind is spiralling in the enclosed yard and it's taking the feathers on a swirling dance. It's freezing cold. The lights and chatter from the house intensify our sense of being in a fairy tale, where we are Cinderella or the Goose Girl, occupied in a mythic chore. We are working fast and hard, gripping little packs of feathers close to the skin and tugging. Sometimes the skin breaks, sometimes the feathers come out clean, but gradually we have more pinky yellow bare skin showing than feather.

The bird is sliding about on our shared lap and the bloodied neck keeps banging against my leg or Clarissa's. I let out a wolf howl to match the grim scene and Clarissa joins in, until we are both spluttering with laughter.

A head pokes around the yard doorway. It's Morwenna.

'Mum, what's going on here?'

Chapter 15

A goat in need of brambles

It's a cold March day with thin cloud and drizzle. It's early morning, before the school run. The daffodils are sturdily upright, generating some cheer, but I am worried about Yamaha. Her pregnancy is causing her problems. She is probably too old for childbearing and is groaning and creaking like a geriatric. I have consulted *Goat Husbandry* and think that her ligaments have become elastic ready for delivery, but in this particular case way too elastic, way too early, and she is clanking and rattling, bone on bone. I feel confident that this is the right diagnosis, but don't know how to help her.

'Stay in the shed', I plead. 'I will bring you food. Let someone else lead the herd for a day or two.' But she looks at me sadly as if to say that without her leadership all hell will break loose. Worse, when I bring in the tempting bowl of oats, she turns away. 'The whole herd must stay in,' I tell her, 'that way you get to rest.' And I bring in extra food for everyone and shut the barn door, firmly.

This is all taking time, but it seems necessary. I run in and see if any teenager might be interested in the current plight of pregnant goats. Our doctor-to-be shows willing, considers my diagnosis with a learned look, and agrees to come foraging with me. We take out a canvas bag and roam the hedgerows, looking for newly sprouting brambles, a goat's almost favourite food. We tweak and pluck all the inch-long sprouts we can find, but the layer of fragile offerings at the bottom of the bag looks piteously few.

I have a memory of some killer brambles in the bluebell copse and we thrash through the long drizzle-wet grass to find them. It's a goodish supply and we pick swiftly, deftly, like native tea-pickers, and then hurry to the goat shed. Yamaha isn't even going to fight off her descendants in their greed to eat my offerings and we will have to distract them, or move them to another shed.

We are running out of time; packed lunch and school-bags need sorting still, but Morwenna grabs Plum and Marigold away, and I offer up the bag of tender shoots, a delicacy to lighten the heart. Yamaha sighs, sniffs them, and looks down at her feet.

I speak quietly and nicely. 'You know you love these.' I take one and offer it close to her soft, nuzzly mouth. No response. I go for force feeding and carefully slip my finger into the gummy pad at the back of her mouth and prise it open and post in the bramble shoot. She ignores the whole procedure for a moment, the complete invalid, and then, forgetting herself, starts to chew. Yes, yes, this is good, and she does me the honour of lowering her head into the bag to eat.

As Plum breaks free and comes over to share in the feast, Yamaha tosses her head in matriarchal indignation. We sneak away.

Peter and the cows

My brother Peter has excused himself from all toil on the understanding that he has an important speech to make and he is far from prepared. He lumbers about the house fussing until I suggest he goes out into the wild woods or the fields to practise aloud, and he agrees.

He is back within the hour, wide-eyed and with a tale to tell.

He had hauled himself over the fence and started drifting along the river edge. He was intensely focused on his speech and was giving it full welly, with hand movements and other gestures and his maximum range of actor voices, when he began to feel he was not alone. He turned to find that Jack's entire herd of bullocks was creeping anxiously behind him.

He waved them away but then they became frisky, prancing hooves, snorting, wide-nostrilled and challenging. He turned and shouted, 'Begone!' and they scattered and came raging in again, eager for this game. 'Away! Away!' Peter lashed out at them, but they loved it and ducked and scampered, willing him to do it again.

So the wise boy used what they offered and started up his prepared speech with all its volume and flourishes magnified. The bullocks listened in stunned silence in a semicircle around him, great brown eyes wide, nostrils wider, hooves firm, and Peter sluiced down on them the full tide of his rhetoric. But as he went on, the bullocks became bored and put their heads down and started grazing.

'My speech is boring,' said Peter, 'I even lost an audience of bullocks.'

Waders

Michael's waders. Fisherman equipment, but for us, essential duck-gathering things. When the ducks are being disobedient and refuse to leave the river, we can use waders as the last resort. Today the ducks are swishing about upriver and quacking a great deal. Something must have spooked them. We are on a rescue mission. Michael dashes into the washhouse and is in too much of a hurry to worry about spiders in his boots. He pulls the waders on and up to his waist, fixes them to a belt, and then his quick march to the yard is a bit like a nappy walk.

Through the gate and into the next field and a firm stride up above the marsh, then he steps into the river above where the truanting ducks are frisking about. The next move is critical. He needs to take a slow, ever so casual, everyday sort of steady pace down the river so the ducks will swim ahead of him, as if it's a traffic queue. If he is too fast, it is seen as threatening, and the ducks will get excited and climb out of the river, scattering across the field. Slow and steady.

I am watching from quite far off, pretending to be thinking about something else, but I am watching. He gets them as far as the fence across the river, where he cannot pass, but they can swim through. Tense moment. Have they forgotten their misbehaviour? Has routine come into play? I call them and Michael gives his duck whistle and they pause.

The first duck climbs out at the watercress spring and starts to waddle up to the yard and soon a line of waddling creatures is returning home.

Hedgelaying

Hedgelaying is an art. You can tell this when you compare the exquisite work on a National Trust property with ours. Nobody wants to see ripped branches, branches at weird angles, or branches tied with bright-blue baler twine and gaps in the hedge.

The intentions and the ambition are good here. We lurch out of the house with high hopes for a brilliant display of good husbandry; we have secateurs and loppers and a cleaver. We have zero experience in doing this, but it seems pretty obvious: you cut part way through a branch and bend it over so that it forms a sideways line and can stop the goats getting through the hedge.

The branches don't much like being cut part way; some of them give up immediately and drop onto the ground cut all the way through. Some of them snap straight back at you without even a scratch to show for it. Snipping with secateurs doesn't do a thing. Nor does lopping. So it's cleaver work.

The first few gashes we make with the cleaver go approximately halfway through and then slide the length of the branch and nearly have Michael cutting his arm off. I am sent in to get a hacksaw and a bow saw. Now the plan is to bend a branch and then at its most bent point to apply the saw. I am to lodge a boot into the hedge and hold firm, Michael is to get above me, without toppling me out, and he is to apply the saw. I am to receive the wrong saw and pass up the right saw without shifting position in the hedge. It is all getting rather difficult.

I suggest bending the branches to a horizontal position and tying them down. Michael frowns. He also considers. It is clearly a very good idea. I am sent to bring out the roll of baler twine, disappointingly bright blue. I reassure Michael that this can be a temporary measure, once the branches have accepted their new position we can cut more easily and later remove the baler twine. He grunts that this temporary measure could be a year.

And so there it is. After an hour of failed sawing and lopping, we have a further two hours of lashing and binding. The goats will snort with derision and push straight through and it looks awful.

Season of cluster flies

Some years they are there in small numbers, crammed into the gaps between a window and an ill-fitting window frame; this year we have a cluster-fly plague. I used to think they were hideous inhabitants of the water-treatment works, fat and lazy from a diet of sewage, then I thought they were farmyard parasites, but gradually I learned they are simply an autumn thing and I had to learn to live with them. But they are the scourge of the world.

When Morwenna opens her window, a bucketful of them fall on the floor and the rest make a discontented hum. She howls in disgust and I tell her to shut the window and I will deal with them later.

I take the vacuum cleaner and post the sucking nozzle up to the window edge and they flow into its dark intestines. Up, round, down and they are gone from the first window and I start on the second. I am a warrior. The flies do not rate the east aspect of the house and the windows on that side are clear.

I think I've finished and I open the upstairs sitting-room window carelessly and lurch back. There are more flies here than I have ever seen. Thick on the ledge outside, crammed in-between the window and the frame, and when I open the window they fall and regroup inside the room. The vacuum cleaner and I respond ferociously and we suck them up from the ceiling, the windows and the floor, until I am aware that the vacuum is becoming sluggish in its suck, so I pause.

I switch it off and peer down the nozzle. I have filled the entire metre and a half of suction tube with crushed flies and there is a sticky, keratin-brown glue with a mix of wings blocking the whole machine. It is stuck like cement. I have eliminated most of the flies and they have eliminated my vacuum cleaner.

A list of jobs

List of gardening jobs for next few days:

Rotovate veg path
Clear and wash the slime from greenhouse
Cut back the laurel
Rake the drive
Gather/rake the trimmings from the hedge in the drive
Prune fruit trees

Clear old brick hedgehog tunnel
Cut fallen apple tree (nice smelling logs)
Cut wild rose from cottage gardens (find strong gloves)
Cut mahonia
Retrieve my wellies from behind the big barn
Check James Grieve apples ripe (think wasps have started on them)
Pick cauliflowers
Shift compost
Clear far parts of wild walk (watch out for primroses)
Cut back hydrangea petiolaris from Owl Cottage
Tie back Japanese wine berry
Trim privet by back gate
Pull down fence by compost
P.S. Remember to buy chainsaw lube
Sprinkle grass seed . . . is it too early?

Oh Lord, really, all this? I think I'd rather curl up with a rug in the conservatory and read *Cold Comfort Farm*.

Raspberries in the rain

There is grey cloud cover and a thin autumn drizzle is soaking everything, and I have to go and pick raspberries. Michael thinks that they are the gift of heaven, but they make me cross. For a start these ones are yellow, pallid and maggoty, and when I start picking, some of them turn to mush in my fingers and others are partly grey with rot. So I pick every one and throw more than half to the ground. No point leaving the dead and dreary ones on the canes.

I resolve to come out before the next ripening fruits have started to become repulsive, knowing that it will be thin autumn drizzle again, and I will be stumbling out in a coat and wellies and a rain hat.

Fruits should be allowed their own season; as I squidge my way over the sodden grass, I reach out and pick a firm, russet-striped, orange-flecked, crimson-flushed James Grieve apple. It is glistening in the wet and it really is a gift of heaven.

Down

Where exactly is down? I should be making pillows and duvets and cushions with all these duck feathers that I have when plucking. So I tug away the pinion feathers, those flying-strength feathers with quills for writing tragedies. And I tug away the waterproof feathers, which are harmless and modest and white. They have some of the downy stuff at the bottom of each, but not that much. I consider these feathers as possible cushion fodder. But where is the down?

When holding a duck, alive or dead, and probing inquisitive fingers into the feathers, you can feel down, marvellously soft, but I had imagined a whole layer of easily gathered down feathers. How are factories that make 100-per-cent pure Canadian goose down duvets doing this thing?

I find some exclusively down feathers. They stick on my fingers and won't shake off. If I gather up more, only a few flutter down into a brown paper bag waiting by my ankles. I try to separate the down feathers from the bigger harder ones. I have many more big ones, and at the bottom of the brown paper bag, I now have enough down to make a 100-per-cent pure Cornish duck down duvet for a mouse.

Hedgerow chutney

We are on a wild picking spree. Everywhere there are things to be picked, so pick, pick, pick. We have baskets of pears and a saucepan full of the most delicious eating apples and I have a good number of bullaces, sweeter than sloes, and of course there are blackberries. We survey a spread of these good things on the kitchen table and discuss their future.

We could have stewed apples or pears for supper, or apples poached with a thin pinking from a handful of blackberries thrown in, but what to do with the bullaces, too sour for most things? Then Michael has the day's bright idea: we will make hedgerow chutney, throwing all the day's pickings into one burst of autumn tastes.

Seven handsome jam jars of chutney. I write beautiful labels for them saying 'Hedgerow Chutney' and line them up on the dresser. Exquisite. A theme starts. This will be the year's Christmas hoard, we will give jam and chutney and apple juice, all home-produced, all tasting of sunshine.

Maud and the branch line

Maud has said she will come for the day to join in picking and weeding and I am so pleased. She knows everything there is to know about growing and bottling from her years on her remote Welsh farm years ago, and I would love to see her. She is coming up from Looe by the branch line and I can see her train as it chugs past.

I run down the drive, over the bridge, along the short bit of road to the siding and I stand on the platform, but there is a problem. The train has stopped in front of the place where the track forks to the right and the driver is struggling with the key. I know this key, Michael has explained umpteen times how it unlocks the track. The conductor is out there too, talking and shaking his head. Nobody on the train seems to have noticed there is a problem, no faces are sticking out of the window, I am alone in my consternation. Maud is on the train and I'm not sure how she is going to get out.

After a longish tussle with the failing key, the driver notices me and calls out, 'Which one of my passengers is yours?' I laugh and ask who or what he has on board and he shouts out, 'I have an old one and a young one.'

'The old one is mine,' I call back.

He looks concerned. 'I'll go and tell her she will have to climb down, cos I'm going to go back to Looe. There is no stopping here as you can see.'

I step down off the platform and walk along the railway track to the second carriage, and the driver climbs up and opens the door where Maud is hovering in the doorway with a clear six-foot drop to manage. 'It's all right my darling,' says the driver. 'I'll go on down and guide your feet to the metal ridges so you can manage it.'

I stand close by and support her back as she climbs down, panting and pink-cheeked with the effort. Standing on the track, we all start laughing, pleased with the success of this particular rescue, and then Maud and I pick our way over to the road.

'Hope the lever and key will be fixed by the end of the day,' Maud calls out. 'I'm not climbing up, you will have to winch me with a rope!'

The borehole

Today the men have come to dig a borehole. It's an emergency measure in case we run out of spring water. We got excited when it arrived and the men started drilling, and then came the awareness that the noise of the diesel engine had stopped. I went out to offer cups of tea and backed off, as the two men were looking hassled and angry and were fiddling about with tools. It looked as if the drill had got jammed or something. I left them for an hour and cautiously crept out again to see if things were looking any better, but the men had disappeared, their van gone. I imagined they had sloped off to get pasties.

By mid-afternoon they are back with more heavy plant, so I brave the situation to ask how things are getting on. It takes a while for the guys to admit that the drilling machinery is stuck and then for them to admit, well, actually, the drill itself is jammed stuck in our drive and they can't pull it out. The support machinery isn't managing it, either.

Then they go off again. And leave the stuck drill and the attached machine.

So now our drive is clogged up with machines and we cannot get past them. It's a bit like having a large guest whom you invited, but now can't persuade to leave.

Cottage cleaning

I am giving my family a life skill: cleaning. Ever since we finished converting the tractor shed into Owl Cottage and the milking parlour into Kettle Cottage, we have doomed ourselves to serious heavy work on Friday afternoons. We abandoned the Saturday arrival time, because it pinioned us to being home all weekend, and now we rush like avenging angels to get both cottages clean and ready by 6 p.m. on Friday evening.

If everyone is home from school by 3.30, we can roar over, carrying prepared loads of clean bedding and a box of cleaning gear. One person is in charge of stripping and making beds, another has to hoover and dust and make the bathroom immaculate, and someone else has to do the kitchen, which means cleaning the cooker as well as the fridge and making sure it is spotless. The last one has to lay the fire, carry in logs, bring over newspaper and cakes and flowers and milk and loo paper, and whizz round making sure that everything is in place and looking spick and span. Then we peel off next door to the second cottage.

The skills training is based on my mother coming to stay at a rented house. We imagine she has arrived at the doorway and is interpreting what she sees. Her eyes flick around the room. 'I notice they didn't bother plumping the cushions,' she says. 'I always think it's nice to iron the tea towels. Ah, a cake.' She squeezes it a little. These thoughts focus the mind. Are there pretty flowers? Is there dust behind a picture frame or on the mantelpiece? She might well have packed a duster on the top of her suitcase, 'just in case'.

So, we have a final look and shut the door with a sigh, then we give ourselves a cup of tea and wait for the first car to arrive in the drive. Michael and I go out and open the car door with a smile and greet the visitors. If there are small children, one of us takes them to pat goats and feed ducks, and the other opens the cottage door and shows a few essential things such as how to light and run the log burner, and to say, 'Please come over if you have any questions.'

Then we rush off and pray that no one ever has any questions.

The Howard Rotavator

Michael has this jewel in his crown of machinery: the Howard Rotavator. He already has a Merrytiller, but it is a mild thing compared to the Hercules of the Howard. The great book of organic gardening says one must double dig, but we rotavate. Once a very long time ago, we hired a rotavator, believing it would transform the field into a smooth tilth, simply by passing this hired beast over the land like a lawn mower.

Michael hung on to its handles in a grim dance of death as it frisked over the stubborn land. Its rotating wheels bounced over the hard earth and took Michael running down the slope. He switched it off and grimly dragged it back to the top and began again, this time pressing it down hard to get a strong grip on the earth. Soon it was stuck in a great hole of its own making and heading fast for Australia.

We learnt in time how to manage a rotavator, but when we were offered an elderly Howard, Michael gasped and nearly passed out on the spot. The king of machines. Instead of a spinning wheel that digs with flailing tines, the Howard moves serenely over the land, thrashing the earth into a fine tilth ready for planting.

Today it has refused to start. Morwenna is standing by in her support role as moll and Michael is red-faced with the effort of again and again pulling the starter cable. I wander by and suggest helpfully that there might be a problem in the fuel line. Michael snarls and Morwenna gives me a warning look. There is a further burst of mad pulling and Michael slumps down, exhausted.

'Is the fuel line clear?' I ask again. 'Is there any fuel inside?' We open the cherry red metal fuel tank with its happy little knob on the top and peer inside. There is fuel, a little, splashing about as we wiggle it, but I can see contaminating flakes of something at the bottom of the tank.

'What's that?' Michael thinks it is rust and we fiddle and wiggle the tank.

We make a bold move. In the dank, dark interior of the big barn there are cans, now obsolete, of thick, black metal-coating paint. I swish the fuel around in the tank and tip it up. Michael and Morwenna share out cleaning the carburettor and poking a fluffy pipe cleaner down the fuel line. I pour little cleansing sips of fuel into the tank and then tip them on the ground. The hens had hoped it was edible, but they shrink back in horror.

When I have cleared all I can, I slop some black paint in the tank. More paint needed. More swishing. I prop it up to dry happy that we have sealed the derusted tank. The other mechanics are ready to reassemble. By mid-afternoon, the kingly beast is ready to go. We stand by in a solemn and reverential group.

Great king, please start.

He roars into life and Michael escorts him into the vegetable garden to make the earth into that fine tilth we need.

Boy in disgrace

Ed is in disgrace and Richard has sent him down to stay with us. We did offer. We have been warned of the causes of his shame, but wisely decide to ignore them and wait for his arrival on the train.

He is chipper, wearing some impossible white jogging pants, a white designer sweatshirt, and a baseball cap back to front, looking absolutely like a London dude gang master, I'm sure. The girls are thrilled, this adds danger and delight to a dull week; they are not at all sure what he has done wrong, and no one is going to ask, in our hearing anyway.

There is lots of listening to vital music with shared headphones, all tucked away in the den.

Then it's half-term and Clarissa and Zafar are down and our house is high with adolescent hormones. Everyone has work projects they should focus on and this afternoon it seems that it is Art. Ed has announced he needs to draw a large landscape and so the devoted team have taken rugs, schoolbooks, and an easel and have absconded up the lane and out onto the top of Jack's field, where there is a flat place.

The easel is up and artwork from Ed is in progress, but the rest of the team seem to be lolling about in admiration. Clarissa and I detect that intervention might be necessary, so we settle ourselves to oversee the project for a little while and reluctantly the team take up their work. After a while we slink away, but halfway down the lane we hear muffled laughter, but do we care? We do not.

Interrupted lunch

We are having a family lunch with Kevin and Mary and their children, Rob and Dig. The smells of lunch are rich and lingering, and our kids are joshing together while we drink beer and chatter. Food on the table, the clatter of cutlery, munching, and an agonised bleat of pain comes loud through the window from the field. It stops us all in our tracks and I throw down my knife and fork and say, 'That's a serious cry, I'll investigate,' and I speed out.

A goat emergency. Marigold has excelled in greed and vaulted over the barbed-wire fence that keeps safe the copse by the river, and she has impaled herself, tearing a horrible L-shaped patch of skin on her udder. I untangle her easily and look at the wound. It's bad enough to worry about; I sprint back and explain what has happened. In the room is a medical team: Kevin and Mary, both GPs, and Morwenna. I ask if they think we could stitch the flap up.

Kevin looks through his doctor's bag in the back of the car and assembles suturing equipment; I bring out diluted warm Dettol and clean scraps of cotton; Morwenna washes her hands. We take our well-fed selves out to the cottage lawn and stake out an operating theatre. Then we lay a tarpaulin on the ground and bring Yamaha in from the field. The other goats all come along too and stand watching from the other side of the fence.

We need strong volunteers to be leg holders. Kevin thinks that we should lie Marigold down, held firmly by the volunteers, with her torn udder within reach of the suturing crowd. Morwenna is to pass equipment, Kevin is to do the stitching, and Mary is to offer advice and oversee things and hold a front leg. Rob and Georgie offer to hold. Michael has two back legs in a vice-like grip and offers one to Dig. I am at Marigold's head, holding her collar.

Astonishingly, Marigold hardly complains about the needle, she is much more offended by being brought down and held, and Kevin gets to work. He manages four or five strategic stitches and swabs again with antiseptic, and we agree that there is no point in attempting to cover the wound. So, Marigold, heal or die, is his instruction. She gets up and shakes herself indignantly and joins her friends on the other side of the fence.

Hot air balloons

Just one time, once in all our years here, and this happens. We are in the vegetable garden, or on our way there, following our busy ant antics, back and forth along our stony drive, and the first we know of it is the sound, a distant whoosh and then quiet. So, of course, we stop and look around for the whoosher and find nothing, but the whoosh comes again and we turn about and then look up to see a hot air balloon, quite low and whooshing to raise its level.

A vast dome of red and yellow stripes. Filling our airspace. Filling it up. We can see the people inside, but we stand still to gaze, open-mouthed. Whoosh, up it goes, but the whoosh is suddenly all around us and there is another one in blue, and one behind it in a whirl of writing and patterns. We can see in a flash that we need to run up the hill, because the wind is drifting them along the valley, but also up and probably over the hill.

We run like mad things, gasping for breath up the steep slope, and throw ourselves over the metal gate and into the field at the highest level. We stop to gaze for the few seconds we have left to see the majestic procession of not three, but seven hot air balloons, dipping over the fringe of hilltop trees, over Jack's farm, and away, away, away, away.

Chapter 16

Midsummer rites

There are three of us gathering suitable magic-inducing plants and supplies for our midsummer rites. We will drive up to the moors and wander until we find the right place to make an evening burnt-offering to the solstice. We are being mysterious, suffusing the harvest of herbs with atmosphere suitable for the three of us as pagan priestesses.

I have the wooden trug and have sprigs of rosemary (for remembrance), a few bay leaves, dry and brittle to release their oily fragrance, and thyme in miserly amounts as my plant is too straggly to spare very much. The girls suggest deadly nightshade, some dried acorns, old apple-tree twigs, thistle fluff, some potato flowers, and we add these to the significant ingredients I have taken from the house. Salt will give off a bright flare and the cinnamon might lend more fragrance, and we have the essential matches and a small bottle of brandy for sipping as we make wishes and mutter incantations.

The late afternoon is still and almost sultry and we set off for the Hurlers. The circle of standing stones is obviously the best place to start our mysteries. Are we too early? There are long shadows cast by the stones, but twilight is going to mean a long wait. We discuss. If we sit hunkered down by the stones waiting for all the dog walkers to leave, the mood will have seeped away.

We face north and walk deeper into the moor.

Hidden treasure

On the lawn by the ha-ha is a piece of ancient carved stone, left from the long-past days when there was a manor here; it is the only sign of any previous splendour, apart from the big granite posts into the yard. Jack pays a visit from the farm above and leans in his customary conversational way on the gate and tells me that there were two big stables here. He says he knows for sure that the old man had buried his box of money somewhere near them, and it was never found after he died.

I look hard at him for the slightest sign of pure Cornish teasing. 'I came here myself and dug, a few years back, when they pulled down the building,' he reassures me. 'Proper fine soil,' he adds, to give credence.

These days it's merely a gateway and grass. But now there is treasure on the brain. We hurry to look at the old deeds and see that there are several buildings marked that no longer exist. Michael gives a cynical laugh, so I have to wait until I am alone and can stab a metal stake into the lawn, hoping for a clunk. We had a few digging moments here before, burying a goat in the extended flower border, or widening the gateway to make an entrance to the holiday cottages, and nothing was found then.

But I am holding onto a thin hope. With each dig, thoughts of treasure flash into my mind, and before I push them away as folklore, I search the ground for a box and scan for a sound of metal.

Slate bench

Georgie and I are on a bluebell-picking mission, treading up through the steep, sloping top orchard with heels firmly planted down to stop sliding on mud and sappy bluebell leaves. They are thick here, densely packed all over the orchard, so we decide to walk up to the edge and take the high route along the beech hedge, and get to the far side before we start picking.

There is clear evidence of badger work up here, little scratchings in the ground exposing roots and a path through the hedge to the fields beyond. The broom and the bluebells are wafting up a heavenly scent and we shimmy under the dangle of catkins, collecting fragments of twig and tiny leaf on our clothing as we push through the more jungly parts of the path.

We need to come through here more often, we decide. This shouldn't be an explorers' route. As we loop back along a flattish ridge in the slope that we have made over the years by firm treading, we agree that we will plant a bench here to give sanctuary and a view out over the farmyard buildings, with their slate roofs splotched with yellow lichen, and the stretch of field down to the river flanked by a thin line of alders.

We are women of action. We fill jars with our bluebells and then start rummaging in the haybarn for some remembered large granite blocks, which are gathering moss where the roof drips rainwater, and then we locate the slate that will serve as the seat. It is a beautiful piece of Delabole slate, once a shelf in the dairy, and is quite long enough for two people to sit in meditation.

We reckon we can manage the tough lifting. The hardest part is the slope at the gate, but we go slowly along the low path, pausing from time to time to shift a grip or catch a breath. We lower one end and prop it against a tree and then we each go back for a granite block, but they are harder to carry and so we load a wheelbarrow and push/pull it together. We need to dig down in the leafy soil with a kick of a heel to level the blocks and then we struggle the slate into position.

Ha! Yes!

We both flop down onto our bench in the bluebells and breathe deep the glory of our view and our cleverness.

Ducks versus crows

In the corner of the shed, the duck that has made a fine nest is brooding a clutch of eggs. She is a bright white and gleams from her dark corner. In the morning when all the ducks rush to feed, she slips out to have a snack and half runs, half flies, down to the river for a morning wash, before returning with a steady dutiful stride to her eggs.

We go into the shed to gather up the random eggs laid carelessly by other ducks and find a crow feasting himself. He has broken and eaten two eggs and is busy pecking into the shell of a third. We quickly shut the door, with him inside and the duck mother sitting on her nest looking alarmed and quacking.

Michael rushes into the feed barn to get some sort of stick and I think better of that and run to fetch the shotgun from the house.

When I return, Michael is in the duck shed and has a dead crow in his hands. He had grabbed the first stick he found, which was a long handle with a curved blade on it, and slashed at the crow and killed it with his first swipe. 'Wow!' I say, rather astonished at this sudden and effective assassination. 'What do we do with the crow?'

Without a moment's hesitation, he strides towards the pig shed and lobs it in. We watch with fascination and horror as the pigs fight to tear it apart. We have 'nature raw in tooth and claw' in front of our very eyes.

Richard and the hedge trimmer

Hurrah, Richard and Judy are down for the weekend, the kids are delighted, we're delighted and there is a Project under way. The seriously overgrown hedge by the meadow needs topping. Michael has set up a dodgy-looking ladder and ropes to tie it to the hardwood and has established thousands of metres of electrical cable extensions to reach the hedge and the hand-held electrical hedge trimmer.

Michael is up the ladder, Richard is looking semi-interested below, and there is a flurry of leaves and twigs crashing around. To keep Richard engaged, Michael has suggested swapping roles, and as I slope off I can hear Michael giving Richard instruction and more instruction and even further instruction. I then hear the whine and growl of the hedge trimmer. A short while later, the growl becomes silence and from the raspberry canes I hear the slamming of car doors and revving of engines and a car driving off up the hill into town.

Twenty minutes later, the car returns and the growl and whirr of the hedge trimmer begins again. Michael joins me in the raspberry canes and explains that despite all instruction, Richard neatly chopped through the electric lead and they nipped up to the garden machinery centre to get a proper repair, and Richard is now re-established on the ladder hard at work.

We busy ourselves weeding, when a second ominous silence from the hedge trimmer suggests all is not well. We think Richard is probably moving the ladder, but there follows a slam of car doors and a revving and a car disappearing up the hill. We remark upon the likelihood of a further repair being necessary.

It is reassuring to hear the whirr begin again when the car returns; the Project is going ahead and we hope Richard is fulfilled and joyous in his work. It's not long before the silence returns and we freeze, straining to hear what is going on. The crunch of footsteps on the gravel, the slam of the car door, the roar of engine, the car climbing the hill, and the thought of the intimacy developing between Richard and the garden-machine guy suddenly reduces us to hysterics. We run to the drive when we hear the car return and drag Richard in for consoling tea and coffee cake.

We all troop out to inspect the Project. There is a very small patch of cut leaves and twigs.

School campout

We are in the middle of a madness of my own devising. This time it is Activity Week at my school and I have blithely suggested that we could have some kids camping out here on the farm doing assorted farm things. My colleague Suki is going to join in.

Twenty-eight girls of eleven, twelve and thirteen are revelling in mud and goat manure and baby animals and we have divided them up into a rotation, so that there are some in the goat shed shovelling manure, another group throwing water about on the pretext of dredging a shallow part of the river, and a third group harvesting raspberries in the vegetable garden.

We have had sessions in the wild wood and a march across the long field to the mud slide to do slipping and sliding on the rope swing, and also much grooming of goats. The whole team is filthy and having the time of their life. I need to get them tamed and clean and fed and bedded down before the next day. We have a barbecue planned, with back-up of beans and brown bread, and I will get them all to walk up into the town to the swimming pool tomorrow, but at bedtime tonight, it's going to be hose water and facecloths.

I try to persuade the girls that it's fine to use the bucket as a loo, which is sectioned off beside the greenhouse with a modesty curtain of an old sail rigged up, but this is clearly a step too far. I make a big deal of praising the plucky few, but there is a steady stream of girls wanting to come into the house to use the downstairs loo. So I recognise defeat on this issue and lay down a pathway of corrugated cardboard and newspaper for the traipsing muddy boots.

Michael sorts out a barbecue for us all and we crouch on bales of hay to eat before going up into the top barn for an evening of improvised entertainment. We have music and silly games and prizes, before everyone climbs into sleeping bags on a random selection of mattresses and camping mats.

I wish I could get back to my own bed. I would have a hot bath and shut out all thoughts of tomorrow.

School campout: home-cooking day

Today's big Activity Week campout project is eat what you grow and the menu is home-made bread, home-made butter, a strawberry cake, salad and raspberries and cream. When I explain we are going to pick and make all of this, there is disbelief. In fact we are cheating a little, as I take some chicken corn and pulverise it into flour to show them the concept and then explain that it would take too long, so we will use a bag of flour that has been ground in the mill at Cotehele using a waterwheel, rather than my Magimix.

We have a group in the kitchen making bread, a group picking salad and raspberries, another group are writing up diaries, ready to swop chores and make butter. The others are gathering eggs and making cakes. Suki and I are in full swing turning chaos to order. The bread is rising under a cloth on the sunny slate bench outside the conservatory and there is a pool of girls lolling about on the lawn.

One or two creep up and ask if its okay to go and stroke the goats some more and suddenly the whole cooking team are ready to fly off and leave us with a half-cooked lunch, while they muck about in the farmyard. We restrain one group to get on with the cake-making and another pair to make butter, but the rest are down the drive before we can protest further and I hear their riotous din as they clank the gate.

The butter-makers are astonished to find themselves whipping bowls of cream until the fat and buttermilk separate, then draining the buttermilk off and washing the remaining butter in cold water. I hand them the old wooden butter hands and toss in salt and they start to squeeze and pat the butter into shapes.

Cake out of the oven, cream whipped, bread risen and baked into big round cobs, butter chilled in the fridge, salad washed, tomatoes picked . . . we are ready for lunch and we spread out our morning's work and call in our muddy tribe to wash. There are proud moments now, 'That's the butter I made, that's our bread, hey, we made the cake,' and these troublesome little girls are suddenly home-cooking queens and they gorge themselves on what they have made for themselves.

Leaving home

It's too much for the man at the railway station when he learns that Morwenna is going away, off to university.

'Oh, you will be terrible sad,' he says unhelpfully, but I appreciate that he has overseen her train journeys for the last six years. 'You look after yourself, little maid, and remember to come home.'

We are all in a fluster. I am trying to persuade her that it is a really good idea to have a sack of potatoes as part of her luggage.

Crows up the chimney

This is going to drive us all crazy: a small cluster of very noisy crows are arguing about who can build a nest in the chimney. There are shrieks and much flying about with twigs stolen from the kindling pile, which are being thrust in. Since we have not been sensible enough to have cappings on any of the chimneys, we are a fair target for a nest, but none of the twigs seem to be holding and there is a growing pile of them on the floor in Georgie's bedroom.

I give her an old pillow to stuff up the chimney and to muffle the crow noises. We decide to be mellow about it and wait to see if the successful nest builder is quiet. But peace is not restored. The successful crow is not a quiet, broody thing at all. She sets up a constant stream of conversation, rattles her nest twigs about, and makes a great nuisance of herself.

Georgie starts complaining that she can't sleep, can't revise for exams, and these remarks to any parent mean action must be taken.

I have a plan. When Georgie finally gets up, we wedge Morwenna's old ghetto blaster right up into the chimney and tune into Radio One, volume max. Utter torment, way too loud, unbearable. Georgie motions to me to switch off. 'What? Why?' She needs to gather up her notes and her books, as she is moving out, heading for the spare room. We go back and switch on again. 'Aaargh!' The noise is too loud for her there, too, so she makes camp in Morwenna's room at the far end of the house.

Michael comes in from the field and asks if we are having a party? He can hear a throbbing beat far down by the river. I worry now about the neighbours over the valley having to listen to this noise for as long as it takes, and I resolve to visit them all later this afternoon if the crows are still in residence.

The party and the crows are hanging in there. It's three days now. We must borrow Liz's chimney-sweeping rods.

Georgie is on sentry duty outside, watching the chimney, while Michael and I have spread a groundsheet on the floor in Georgie's room, removed the music system, and are threading together the sweeping rods, each length of a few metres going higher and higher until the rods meet resistance. I lean out of the window to yell to Georgie that we have probably made contact, so she should watch carefully.

Michael heaves on the rod to force it up, heaves, heaves, and Madam Crow and her belongings are shot into the air in a noisy clatter of falling twigs.

She is outraged, we are delighted, Georgie is whooping down on the lawn and once more, quiet can settle in the valley.

A party for Morwenna

She rings only three days away from her birthday at the end of May, battered by the strain of the exams now over for the year. 'There isn't a party anywhere this weekend,' she says softly. 'It's my important birthday . . .' and she lets the words hang in the air. She had already told us not to even think of what she needed for a party, as there was sure to be something happening.

Quick as a flash, I say, 'Don't worry, I'll fix you a party for Saturday, just let me know numbers. I'll do the rest.'

Georgie understood the content of this conversation as she slumped over her breakfast. A party, Morwenna needs a party and we both roll our eyes. Typical. But crikey, what can I fix? Games and jelly? Hide and seek in a forest?

'Arabian nights,' says Georgie, through a mouthful of Coco Pops.

I gaze at her, transfixed. 'Brilliant. What do we need?'

Georgie shrugs and says, 'Hmm, harem pants and belly dancing.' I rush upstairs to the spare-room cupboard, the home to three drawers full of dressing up stuff and pull out several lengths of Arabian-looking fabric and one pair of harem pants, lots of diaphanous scarves, lengths of muslin, some fancy belts, bling necklaces, some other baggy trousers and an embroidered kaftan, and I shove it all into a basket. Georgie then offers some floaty stuff of her own and I assemble a pile of floor cushions.

My mind is working fast. I can drive up on Friday night, complete with food, music, masks, Turkish slippers, cushions, and further costumes, which I can make tonight. I remember that in the school Drama cupboard there are two pairs of glittery harem pants and matching scarves with stitched coins on the edges and, most importantly, a hula hoop from which I can suspend all these lengths of fabric to make a tent. I will bring these home tonight.

I tell all this to Georgie, now on her toast, and she nods in agreement until I ask her about the belly-dancing music. She has nothing, I have nothing. 'You'll have to buy that,' she says, 'probably in the motorway service station.'

My mind now spins to the food, and Georgie has disappeared to get dressed and start her A-level revision for the day. I've lost her.

Turkish delight, falafels, stuffed vine leaves, spicy couscous with pomegranate, a Moroccan lamb stew, pitta and feta, grapes. I completely forget that Morwenna hates lamb. I bring ingredients out of the freezer and store cupboard and survey my hoard. A shopping trip is needed on the way home from school. Maybe I will find someone with belly-dancing music. We have a tambourine. I will need to do some hasty work with the sewing machine.

I make a quick list in my book of things to take, things to buy, things to do tonight, things I must tell Morwenna. If I can get her to leave me alone in her house on Saturday afternoon, I can assemble my Arabian nights room, and get all the dressing-up stuff laid out in her bedroom for her guests to pick through and adorn themselves.

I ring her and say, 'Can you limit the numbers to ten and tell everyone to arrive at your house at 7 p.m. with clean pants on?'

Morwenna says, 'Oh, wow.'

Woo in the snare

I am looking for Woo. This sleek black, half-Burmese cat never ever misses breakfast, but today only Ezekiel lined up for cat stars. The two are usually found skulking around together, mostly because Woo has taught E everything there is to know, mouse-hunting, waiting patiently by a molehill, how to go too far away and still get back home. So where is she?

I take a long, searching route from the house, down into the meadow and along the river edge. It's beautiful. The river is running fast and the hemlock is knee-high and I stomp on it as I go as a productive part of my walk. I keep an eye out for kingfishers and any stuff that may have snagged on the alders on the riverside. As I go, I call out my Woo noise, but there is no reply.

I loop up into the yard and as the gate clangs shut, metal on metal, I hear a faint Woo call. I stand still and call her and get a thin reply. I can see her straight away, a black bundle suspended midway in the fence. It takes a moment to work out what I'm seeing, but it's Woo caught in the snare Colin put up to catch the fox.

I approach her slowly and talk to her. The wire has pulled tight across her chest, but this has prevented struggling or strangling. I'm not sure how she will react to any attempts to free her. In a panic, will she bite or scratch? Will she remember that I am her friend? I keep talking and reach out my cupped hand to take the weight of her body; she starts to purr.

Oh, Woo.

With the other hand, I can work the wire loose enough to push her head back through the loop and I scoop her up and hold her close. She purrs and purrs and as I stroke her, I search for injury, but find none.

Haymaking

There is a mad harvest mood in the hot and humid valley. You can see tractors lugging their machinery along the road to St Keyne, hear them rumbling through the night, see their headlights swooping a beam across the fields, and smell the dust and fragrance of drying hay in the air every minute of the day.

We are waiting. The hay in the big field is cut and Colin came down from his hilltop eyrie with the tractor and its enormous metal spike attachment to whisk the flat piles of grass into a fluffed-up line to dry the underside. That was yesterday.

Sometimes the hay needs two or three turns, but it is so hot that everything is desiccating as we breathe. It could be gathered up and baled today; it should be gathered up, because the increasing humidity signals rain soon, a destructive, mouldering rain. So, again we are waiting with a clenched tension, and too frequent glimpses at the sky, or checks on the distant noise of a maybe approaching tractor.

It's pointless waiting, as we all have our lives to get on with. Baling and bringing in the hay will happen, if and when it happens, and I have to get myself and Georgie to school.

We swelter all day in classrooms and drive home sweaty and annoyed with traffic now trebled with lines of holiday makers and caravans. We know it's a pointless hurry, but it's how things are, we have a powerful need to be there, drinking tea and staring at the field, anxious to start bringing in the harvest.

When we get home, we find Michael in the drive loading up sailing gear. He shouts out, 'I'm off racing at Torpoint, but as you can see, Jack has baled up the hay. I think it might well rain later on, so can you two gather in the bales?'

And he is off, leaving us dumbfounded. I try to flatten out the panicky rage that has filled me up entirely. Sailing? Really? Usually we have one person on the tractor, one bringing the bales into tidy clusters, and one more heaving them up onto a trailer. But now it looks like no tractor or trailer and it is up to one teenage girl and a feeble mother. Blast him.

Georgie and I decide we can do it by loading the little Suzuki van with four bales at a time, shoving them in and both working together to lift one to the top layer. Georgie seems quite chuffed to be allocated driving the van, since she has had only two driving lessons. I am wondering if our stamina will last for a hundred and fifty bales.

Georgie manages the van fine and we are working well as a team, but the bales are heavy. We have both got gloves on, so the twine doesn't cut, and we are adopting a cowboy-like stance as we heft the bale onto thigh and then up into the van. We laugh at our clumsy attempts, we laugh as Georgie drives into an unseen bale, and we laugh as we back the van into the hay store and heave bales around into the base layer of a castle, which we know we aren't going to manage.

We are very hot and little bits of hay stick to our dripping necks and faces, creeping into cleavages. We take a break for lemonade and take stock of our labours. Six loads of four and an hour has passed. Hmm. We clearly have a problem and we make dark mutterings against those whose priorities are clearly wrong and should not be sailing.

But there is a tractor noise and a shout at the edge of the field. We squint to see who or what and move fast towards the gate. Standing and grinning at us, complete with checked shirt, broad shoulders, and a tractor and empty flatbed trailer stands Andrew, one of Michael's handsome sixth-formers, whose farm is in the next village.

'Sir said you might want a hand,' he says shyly, looking at Georgie.

In an instant, Georgie abandons the van and the three of us are hauling bales up and stacking them, onto the trailer, onto the stack, trailer, stack, trailer, stack until the field is stripped clean. It's suddenly fun and fast and there is the delicious race as the first small spats of rain start to fall and we have just beaten them. Our harvest is in.

Much later Michael returns with a sly smile, and says, 'I thought you might like a bit of help.'

A garden for Clarissa

My friend is dead, my soulmate, my springtime friend has gone. Hers has always been the greatest joy in spring things, in chicks and daffodils, muffled-up walks, a bright new yellow outfit for Easter, in all things bright and beautiful. So, armed with buckets of compost, bulbs, small shrubs, gloves and some thoughtful but keen helpers, we are making Clarissa a garden up on a sundrenched bank that looks over the cottage courtyard and lies alongside our woodland path.

The earth is a resistant tangle of roots and we work hard at getting holes deep enough to nourish with compost, before packing in our pot-grown camellias. The bulbs are easier and the girls have my father's old bulb-planting tool, cone-shaped and with a handle for forcing down the cutter, which pulls out a plug of soil. A daffodil bulb goes in and the plug is replaced and pressed down. I want a mass of daffodils and then the camellias will take over the flowering, enough to last all spring before the bluebells take over.

A mass. A million nodding heads. A small sunshiny forever cry of love.

Workshop

We made a long and essential workshop for Michael, and it's partly for useful man-work and also for storing tools and a multitude of junk. To stand in the doorway and look down the narrow pathway that separates the piles of things is to gaze into a whole world of miscellany.

Boxes of wine and racks of wine, a drum of standing pieces of dowelling and sticks and fishing nets, racks wall-mounted for tools, a heavy-duty wooden workbench laden with bits of stuff, a lower shelf with glass rods, a big box of cut pieces of matchbox-size wood, another box of chainsaw stuff, a bench-mounted anvil, a free-standing anvil, a drill, a vice, oil lanterns, wire looped and hanging from the rafters, Barbie body parts, a fox snare, a plastic box for ducklings and an infra-red lamp, a polystyrene incubator, some parcels of Victorian coloured glazing, a chair I'm meant to be reseating in plaited strips of fabric, shelves with paint and brushes and rollers, bike spares, a chest freezer, a 12-volt battery and charger, a drooping inflatable gorilla, a Seagull outboard, an electric-fencing battery and charger, a crowbar.

If we ever we sell this place . . . Oh, my goodness.

Chapter 17

Rain and flooding

Rain and rain and rain, all night drumming on windows, all yesterday, maybe all today. When I look out of the window into a drab sky, I almost sense the flooding before I see it. The river has sat up and climbed over the banks, over the knotted roots of the alders, and marched purposefully into the field, carrying twigs and rubbish and a visible current. Where the current isn't tugging, the water reflects the sky and silvers a little.

At first the ducks are ecstatic, they are lords of their watery world and they dive and swirl in their new empire. It's a short-lived rule; they are suddenly upset and indignant at the arrival of some barnacle geese, who have flown over us in a vee formation and, seeing a great stretch of water, have circled and landed amongst the ducks. Our own geese ignore the new arrivals and sail like galleons down the valley.

The hen who hatched duck eggs is beside herself with anxiety as her surrogate children twirl off into the water; all she can do is shout to them from a dank patch of nettles.

By midday the ducks and all the geese are in the next field, the bedraggled ducklings are warming up under their foster mother and a heron has taken ownership of our flooded meadow, wading disdainfully over the drowning buttercups. The water level is shrinking, the rain has stopped.

I climb over the gate, checking the fence for river flotsam, and try to heave away some long branches that are stuck, forced by the current. It is going to be a job with waders on, thigh-deep in the river. I drag a long branch worth burning up to the log store and then think myself into a reward of a crumpet and Marmite.

Naming our land

It is a known fact that Michael and I disagree on almost everything, and so why not disagree on where we actually live, what we own here, a piece of land with several dwellings on it.

We surely have a small farm. We have a farmhouse, oldish and stone-built, on the site of an ancient manor and presumably once an attendant mill, given the lie of the flat land next to the river. We have a courtyard of barns, a big stone-built barn with stables below and a huge space above with a gap along the top for pushing hay down to the mangers underneath, and several smaller barns once set up for livestock, with feeding troughs and mangers and drainage channels. A farm, surely. But we have only a small acreage, as most of the land has been passed on to the neighbouring farms by the Duchy of Cornwall. We have seven acres of orchard and mixed woodland and meadow and garden.

Michael says we have a smallholding. This also is technically correct. We also have an enormous menagerie of creatures, so maybe we are a zoo. And then there are the holiday cottages plus playground, so maybe we are a holiday complex.

So what are we? Are we teachers, since that is now what is using up way too much of our time and energy, or are we farmers? Or are we smallholders? I don't feel like a smallholder, it sounds like a doll holding a chicken. To say we are farmers feels like a boast of umpteen acres; to say we are running a holiday complex sounds too wordy and smacks of Butlin's. I'm definitely not a zookeeper.

We are experts in crisis management.

Mowing frenzy

Friday afternoon, after school, everyone home, and is there to be mowing tonight or tomorrow morning? 'Get it done,' says Michael. 'Have a gin and tonic,' say I, but before he has heard me, he has sprinted off to the barn, that good old barn stuffed with feed bins, assorted tools and barrows, and a fleet of lawn mowers.

So Georgie and I stroll out and survey our choice of machines. Michael is starting them up one by one and I take the old green mower with the number plate and set off down the drive, scattering small stones and taking a sharp turn down past the two holiday cottages, and I start on our lawn. I know that Georgie will have been offered and refused the blue mower for the courtyard lawn and that Michael will do that once he has got Georgie started on the ride-on.

She bursts into the garden at breakneck speed and starts a loop in an anticlockwise direction. I hastily turn to follow in the same direction, for safety's sake, and start cutting a looping stripe just overlapping hers. Round along the bramble hedge, past the dark buddleias, nip around the stunted yew trees, along the ha-ha, up and dodge under the purple prickly bush that grabs your hair as you pass, and make a perilous sideways swerve along the steep slope under the house; then slither around the big yew tree, up past the conservatory door and along the herbaceous border up to the old privy, past the impenetrable thicket of the pink rose and finally along the Cornish hedge. A lap completed.

Georgie yells that she is going to overtake me, so I cut across under the apple and pear tree island to make a half lap.

After about three laps, some full, some half, depending on how terrifyingly close Georgie is getting with her faster and more powerful machine, Michael joins us with the blue mower and chugs along contentedly on half laps. Now he is in danger of colliding with me. Ten laps, fifteen laps and we are nearly done. There are strange little tufts remaining where skill and manoeuvrability have failed, so I vroom over to sort them out. Finished in record time.

Michael clearly can't believe this winning performance and glances at his watch; sensing a moment of possibility, he suggests that we should tackle the drive and the driveway lawn. Georgie and I give him a look and roar off down the drive to the lawnmower hostel.

We have a gin and tonic in mind.

Sleeping

There isn't a sound. I almost feel that something is wrong, the world has ended and I hadn't realised. I strain my senses to see or hear or feel that life is continuing. Ah, I can hear a distant mechanical hum, maybe a car far off, but it fades and it isn't enough to reassure me. I cross through the conservatory, which is awash with chaos, mess, muddle, yard coats, discarded wellies, and a meagre harvest of apples and carrots brought in for cooking. In the midst of all this lies a stretched-out strip of fur that is E snoozing.

I crouch down and stroke him and he gives a half-hearted stretch and starts to purr. I'm pleased that some-one is alive. I ask him to come to the yard with me, but he is too sleepy, so I set off alone in the silence.

The gate offers its familiar squeal, but where are our living creatures? The silence makes me creep into the grassy courtyard and I can see no sign of life. Through the next gate, I peer into the duck shed. Nothing. Into the goat shed. Nothing. Chickens, surely, somewhere? Or has a fox come and scared everything? In the next shed I find a hen on the shelf of the manger and she is laying an egg, or planning one, and I could almost hug her with relief, but she gives me a stern look. In the pig shed, it takes me a moment to adjust to the darker shadows and at first I can't see pigs either. But there are piggish shapes in the loose hay and then I see a trotter. Alive, then.

I open the field gate and stare out. All is well, it is just that everything is asleep, the herd of goats stretched out in a sleepy pile, some hens basking in the dust, a cow curled up with drooping head. Asleep, everyone, silent. It's that midday snooze time.

Plastic sheeting for swallows

We need to provide a laundry room for our two holiday cottages. I have been puzzling how to arrange this and have a eureka moment.

Michael is willing to follow my plan and obligingly divides the milking shed into two with large panels of marine plywood and struts. We make a hole at one end through which a series of water hoses can come and go. The shortened shed still has a shelf on which goats can stand to eat and be milked, and the milker still has the cork-topped bathroom stool on which to sit.

But the other half . . . a laundry room! A washing machine stands on a wooden pallet, because the floor cannot be trusted to remain dry in rainy weather, and so too does its friend, the tumble dryer. I paint the rough stone wall with lime whitewash and put out a welcoming assortment of laundry baskets, pegs, drying rack and a few useful cloths.

It all looks very good . . . but I have forgotten the swallows. They like to nest here in the roof. When I go to gloat and check how wonderful the little laundry is, I find copious small droppings, fallen or abandoned bits of nest rubbish, and scattered feathers. This will not impress my washing visitors.

Michael gives a weary look when I tell him I have a further plan, but he comes with me to the builders' yard and we buy transparent plastic membrane to make a false ceiling. Michael teeters on a ladder with the plastic and I balance on the milking stool hammering in nails. He holds and tells me where to hammer. We have to swop jobs. It sags and hangs a little in the middle, but there is room for the swallows to swoop in.

Dave and the cockerel

Every Friday when we go into the cottages to do a speedy clean, we glance at the visitors' book and crowd round to judge the comments. Most weeks someone complains about being woken up by the cockerels crowing from 4 a.m. onwards. Currently we have a quarrelsome batch of cockerels who have started up gang warfare, so I have decided to cull some, but a friend at work has told me she would like a cockerel.

Really? I am astonished but name a small price and pick up the most aggressive of the fighting boys, then put him in a box and drive him to work with me. I am to deliver him at the end of the day to his new home and to a cluster of new wives. I am leaving him all day in the car (with, of course, water and an open window).

I pick up Georgie from school and she asks me to give her new boyfriend a lift home, so she hops in the front and he sits behind me in the back seat. We set off towards his house and some scratching can be heard from the boot. Dave's eyes open wide with alarm. I catch his glance in the rear-view mirror. I say nothing. There is more scratching.

He speaks up, 'I think I can hear something moving in the back of the car.'

'Really?' I say.

So now he is sure I have vermin in there. Suddenly the cockerel gives out a monstrous cry and Dave rears up in terror, going grey in the face. 'Can you let me out here at the bottom of the road?' he says and scrambles out.

'What are you doing?' Georgie asks me.

'A cockerel in a box in the back,' I explain.

'Mum, you are absolutely horrible; poor Dave.'

Hemlock

Down by the river the grasses and weeds are so high that the goats have started to ignore them, and we can no longer see the river burbling past. I am dressed for attack. I have wellies and the strimmer. I don't need the visor, as my specs will keep flying bits away from my eyes.

Michael always enjoys assisting with the preparation for a project like this, ordering his army of one, and he has prepared the strimmer with a top-up of fuel and he is trying to get me to wear a pair of gloves. I refuse, too hampering. He helps with the pull-cord starter and I cross the yard sounding worse than a chainsaw. I feel sorry for our neighbours on this Saturday morning.

I leave the watercress by the spring and begin my devastation by the first alder trees, slicing through a bank of nettles as high as my thigh. I'm careful to make sweeping cuts to fell the stems away from me and I don't bury the cutting head in too deep, afraid it might tangle and strangle itself with the fibres of the weeds.

The weight of the strimmer is borne by the thick shoulder strap and my sweeping motion is working well as I progress along the bank, through nettles and a forest of hemlock. It is annoying to see the hemlock has grown so high and lush, as it is supposed to be poisonous to cattle, but I think only the roots are dangerous. Anyway, the goats haven't browsed it, so I need to cut it.

Michael comes down to gesture to me if I am managing okay and I give him the thumbs-up.

My shoulders are aching and my hands and wrists are fizzing with the vibration of the motor, so I am glad when the fuel runs out and I can stop. I haven't quite finished, but that is enough for today. In the conservatory, I see I need to strip off my clothing, which is entirely covered in little flecks of green vegetation. My bare arms and face are also flecked with green bits that won't splash off in the sink. I am in need of a proper shower to sort out my green hair.

In the morning, I'm surprised to see in the mirror that my face, hands and arms are covered in shiny red blisters, sore to touch. I look awful. I wonder if I can even go to school. My unsympathetic family hoot with laughter.

I have a bad case of burn from the sap of the hemlock and I remain disfigured for over a week. My pupils and colleagues treat me as if I carry the plague.

Teenage shoes

Michael is the first one down this morning and, with his cereal spoon raised to his mouth, he nods significantly in the direction of the kitchen door. I open it and there on the mat is an unknown and enormous pair of shoes. I look back at him with an answering question mark and put the kettle on.

The girls were out last night, we didn't hear anyone coming in, and we certainly aren't about to go upstairs and do a roll call. We go out into the yard for milking and dealing with the animals and a prolonged session of cutting goat hooves.

It's eleven by the time we are done and ready for coffee. At the kitchen table there is a collection of bleary-looking teenagers picking over whatever happened last night while clubbing in Plymouth. They are full of hair-raising tales that parents don't want to hear, but we hang about sipping coffee and absorbing it all. John, the son of our friend, is the owner of the large shoes and he thanks us with exquisite politeness for allowing him to stay the night in the spare room; he had missed every chance of a train or taxi back home to Morval.

'That's fine, John,' we say, and it obviously is fine, as the frequent extra shoes over the weekends ahead reveal. We are a very fine stopover.

Not a beach day after all

Heaven here, hot, sun teasing and tweaking, the call to the beach and the sea and to lie on the ground is deafening, far far louder than the call to weed beans or mow the lawn or buy goat wormer from the vet, far louder even than the call to pick flowers.

The phone rings. I answer. The washing machine signals a command to be emptied and the contents hung out. I comply, but must be quick before the need to run away to the sea has been overtaken by the duties of the day. I ignore the second phone summons. And a third. I scramble for a bag and shove in swimming costume, sarong and an apple and run up the steps and escape.

Halfway down the drive Jack's tractor blocks my escape. He is in loud mode with his brother Robert and there are bullocks roaming where they shouldn't. My help is needed with the herding job. We seem to be managing until a frenzied group of three break through and run up the hill. I have to get back in the car and reverse up and block the way so they don't come into our yard. But they get there before me.

The gate is shut, they don't know where to go, they lurch and rush and I pray they won't rush down the steps to the garden but they are too buzzed up for that, thank goodness, and they go back down the hill.

Again I hear the phone ringing. Change of plan then. I tell myself that I can lie in the grass after I have done the weeding.

More vine work

It's hot summer and the vine has produced a tangle of huge sappy leaves and twisting shoots. I need to cut them away to let the bunches of grapes get more sun and more nourishment from the stem.

The cutting away is easy enough, I let great branches fall to the ground, but I am troubled by the bunches. The grapes have swollen and are pressing tightly on each other and this will lead to rot. I need to thin them out. I sigh a bit and prepare a system that lets me walk about under them without hopping up and down on a ladder.

I set up several chairs and two folding stepladders and move between them to stand underneath the vine. Holding a bunch in one hand, I juggle and jiggle it to see if I can loosen the grapes and then start pinching out a few grapes at a time until the bunch can move freely. The next bunch has already got into trouble and there are several grapes mouldy from being pressed too tightly together. I thin out that bunch more ferociously.

This vine that I love has thirty-seven fine, fat bunches on it and by the time I have jiggled and thinned each of the bunches, I have neck ache, shoulder ache and backache, and a scream from my pinching thumb and finger. But the deed is done and I have saved the crop. In a few weeks, the first of the grapes will be flushing purple and soon after that the whole lot will be sweet, black, warm, and utterly delicious.

Mangetout and picking

If I get up quickly, I can do some picking before breakfast. There are mangetout calling out for picking and if we don't, they will grow fat and leathery, so I must hear their cries. The whole vegetable garden is shrieking. There are little beetroots ready, but they can wait till the weekend. There are broad beans to pick, but I hope I got most of them the day before yesterday; there are gooseberries and the start of the currant harvest and some raspberries, too. And the spinach. And some new potatoes. But I can't do all that before breakfast.

When we get back from school? I have coursework meetings and then enough marking to pin me to the table till dark. So, I set a target for the evening. Do ten coursework markings, then dash out for crisis picking of only one crop; come back in for more marking and a cup of tea, and then go out for another crop. Someone might join in?

It is the time of produce. It is always going to clash with the time of coursework. But we have the long, long evenings and I can pick by the light of the silvery moon.

Hammock break

Aargh, the shame, the embarrassment, the damned inconvenience. There is my big party tonight, with a live band, dancing and eating in a marquee and a bouncy castle, and I am badly injured.

Morwenna and I are basking in the conservatory hammock, two happy things in a moment of quiet reverie, when the cord snaps and drops us with a thud onto the granite floor. Shocked silence and agony, then Morwenna in full doctor mode manages to gasp, 'Lie still, Mum. Can you wriggle your toes?'

It takes a moment before I can say, 'Not sure, what about you?' And we both lie there, still, but toes wriggling.

Georgie, of course, laughs to see us sprawled out, but is given serious instructions from the doc. 'Get frozen anything and wrap it in a tea towel.' We must each have a frozen parcel on our possibly broken backs. Georgie pulls down our jeans and underwear and puts bags of frozen peas on our agonised bottoms. Thus we lie, in a mix of regret, pain and despair.

Suddenly there is a whirr of tyres on the drive. A slam. And footsteps.

Georgie is quick to leap out to intercept our visitors. It is my goddaughter Kitty bringing her new boyfriend to meet me in advance of tonight's party. There is a moment of hesitation from Georgie in the doorway, but Rupert comes forward without a flicker and lies on the floor next to me to introduce himself.

Georgie and Kitty attend to covering cloths and servings of ibuprofen for the frozen-pea casualties lying on the floor, and Morwenna and Rupert and I chat as if there is nothing awry. But I am full of dread for tonight. Can I even walk?

We are the girls who love to dance, to whirl, to leap on the bouncy castle, and it's my fiftieth birthday.

In praise of ibuprofen

Not much can get in the way of proper party lovers. How wonderful to gather all those friends and laugh and dance. Raise a toast to the power of painkillers. And champagne. And sheer adrenalin.

Pressure hosing

Now why am I doing this? Is it because we have a shiny new, blue mechanical toy with which I want to play, or is it because I need to tidy up our farmyard? Easing out the new pressure washer from its box and carting it down the drive in a wheelbarrow has raised a level of excitement. I have even put on my old blue boiler suit. Michael has given me a look that says many things, but mostly, 'Oh Lord, what is she up to now?' To avoid further interference, I am now doomed to read the instructions on the box, which is not a thing I much like.

We need an extension cable, so the wheelbarrow and I go to the feed barn to collect the heavy roll, which has a broomstick through the central hole. With the broomstick suspended on the handles of the wheelbarrow, I can unroll any length of cable. I plug it in to the switch in the goat-milking shed and then connect the hose. This will be the limiting factor: I have only a few metres of hose.

I grab the milking stool and stand with one foot on it and the other raised onto the wall of the pigsty and I press the trigger on the nozzle. Nothing happens. I climb down and switch the machine on this time and then step up again. The sharp jet of water shoots up into the roof beams, which are heavily swathed in dusty cobwebs, and they start to hang down in thick clumps and then drop to the floor. I move the jet steadily and methodically along the beam, reaching into the clusters of web and driving more and more down. I have soggy lumps falling on me and water running down my arm, but this is thrilling stuff and I am not stopping yet.

I climb down and reposition the milking stool and attack the other slope of the roof.

The milking shed now has sparkling walls and is almost ankle-deep in gunge below, so I take up a back position and drive out the clumps through the door with my watery jet.

It's power and transformation. I lug the machine into the pigsty, but now I require a longer hose. I will need Michael to sort it. I brace myself against the cold wind, which is going to freeze me as I dart back to find him and I plead. It's worth it.

Fencing

When I zoom into the drive after school, I can see Fernley's van parked by the hay barn and I am jubilant. At last, at last, he has come to put up the fence around the piece of land we have bought from the Duchy of Cornwall: Jack's land that is being sold or parcelled off now that his tenancy has ended. It's a few acres only and includes the wonderful hedge halfway up the hill.

I have been thrilled to think of the extension to my run around the perimeters of our land; now I will be able to break through the end of the nut walk, or run along the far side of the hedge along a new pathway and swoop down the slope of the big field and curve back in along the riverbank. I will be able to curl up, snugged into the bank where the primroses grow so thick by the hedge, and gaze down to the bridge. E and Woo will be able to pounce out at me from a brand-new angle.

I cut through the vegetable garden and find Fernley coming down from the field with his shovel over his shoulder. He takes me back up to look at what he has done, and I nearly slump down with shock. We do have a fence made of fine upstanding poles, with sheep netting stretched between them and a double row of barbed-wire on the top, but it's in the wrong place. I catch my breath.

'Fernley,' I say, 'we agreed where the fence was going. It was to be a metre from the hedge to allow access to the hedgerow trees and bushes for cutting back and pruning. We need to be able to walk behind the hedge.'

He looks caught and worried. 'Yes,' he replies edgily, 'but the guy who is about to buy the old Duchy farm was here and he told me to put the fence just here.'

I don't need to remind Fernley that it is we who are paying him, and with whom he has a contract, but I can tell by his look he had been woefully bullied. I look up at the sky, holding back tears of rage, frustration, dashed hopes, and grief. A new neighbour has just stabbed me, before he has even moved in and for less than a metre of land.

Georgie's party

It is almost midnight. I step out of the wild party scene and steady myself down the vertiginous ladder from the top floor of the big barn and I creep outside. It is a cold March night, but there are a million stars shining, and it is refreshing to gulp in the cool still air. How have we managed the luck of a huge birthday party with no rain? Ceilidh music is thrumming all around the courtyard of buildings and people are drifting about; I can see the tiny glow of a cigarette end, or the black shadows of people as they drift past the lanterns and candles.

We have polkaed and twirled and jigged and partnered and spun and shrieked with laughter. Morwenna and Rory each swayed and twisted, facing each other out in 'High Noon' challenge. Georgie has insisted we are all dressed in more-or-less cowboy style and various toy guns and hats are scattered about.

I walk into the little stable stalls that have briefly become sitting areas and consider safety. Yes, the floor is moving, but the adjustable steel Acrow prop is holding it all up. The slippery hardboard floor covering is keeping dust and spiders down. No one has fallen out of the open doorway upstairs. Michael and I should leave while we are ahead.

Bluebells

If I stand and look out of the bathroom window, I can see into the top triangle, an acre orchard of loved trees, the Leaning Tree, Mare's apple, the Sturmer, Pa's Bramley and countless hazels, pears, and cider apple trees. But I'm looking for bluebells. In the first week of May, I should be seeing a blue flush over the grass and a million bluebells should be pushing their way into bloom from their sappy pointy leaves.

Kitty has asked me to pick bluebells and apple blossom for her wedding and I paced through the triangle and found not even one stalk of bluebell; ridiculous, we always have bluebells in fragrant plenty. Her wedding is too early for bluebells. Dean says she has seen a flush of them on the coastal path near Fowey and that Meriel has crab-apple blossom in her orchard. So Meriel and I pick great boughs of the blossom, pinks and white, and I store them in old milk churns in the washhouse.

The day before the wedding I set off alone along the cliff path, gorse filling the air with the smell of coconut, scanning the slopes for any hint of blue. Bracken fronds are unfurling and shimmering with silver fluff and I can see new bramble shoots, early daisies and violets, but no blue. Just as I am turning back, a hint of blue catches my eye higher up on the cliff slope. I push my way through the knee-high bracken and climb up.

Yes, a few bluebells, enough maybe for a wedding day. I slide my fingers down the stem of first one and then others and snap the stems at the base. The slide and wriggle of a thick, brownish adder only inches from my fingers makes me leap and then pause to realise what I have seen. Reason tells me I should leave well alone, but I need more bluebells; it was just one snake and it has gone, so I resume my slide and snap, slide and snap, gathering great handfuls of bluebells very quickly.

A few metres up the bank, I surprise two more adders curled together and they shoot forward; I shoot back. I have a sudden panicky vision, the whole picture in front of my eyes: the sharp, venomous bite from each, my collapse in the high bracken, invisible to all walkers, my slow death, the bluebells limp and dying like me, the wedding cancelled . . . I need to leave.

I stumble down to the path with my trophy bag of bluebells and race home. The bluebells and apple blossom are heavenly in the church as Kitty marries Rupert.

Chapter 18

Watching badgers

We know there are badgers at work, we see their scratchings on the lawn, while in the field and in the woods, we can see their smoothed paths up the banks with occasional claw marks. In the woods by Trussel bridge we see a badger city; there are abandoned setts, active ones with discarded leafy bedding, and a little way off a series of latrines.

Judy and I decide to watch them and we set off into the twilight each with a cushion, a torch, a blanket, and warm clothes. Behind an old boundary wall fallen to rubble, we crouch out of sight and downwind. It grows dark and darker and we sit silent and don't move a muscle. My ears are straining to interpret every rustle or sigh of the woods, but I don't know what badger sounds to expect. Surely a grunt, some sort of badger chat? I imagine they know we are there, and they are all huddled in their setts, miserably wondering what fate is in store for them.

I can't see my watch. My body seems to have turned to stone. I daren't look at Judy in case I start to giggle. I clamp my jaws tight and think badger thoughts, until I realise I have fallen asleep. How long?

I turn to Judy and whisper, 'What can you see?' and she replies, 'Nothing, nothing. Let's go home.'

Our limbs are stiffened in agony as we stumble away. I picture the badgers giving a collective shout of joy, running out to hunt for worms, and then laughing at us on the farmhouse lawn below our sleeping bodies.

Memorial planting

I have too many dead people in my life, out of my life. I ask Michael if I am being macabre, but I want to plant a tree that will flower in January as a memorial to Chris. But Michael is so choked up by Chris's death that he won't answer and I drive off and buy a mahonia that will be heady with fragrant flower heads when nothing much else is, and I dig it in with as much solemnity and sadness as if I was digging a grave. So the mahonia is there, nourished by tears.

I plant more; secret bushes and trees to commemorate. For Clarissa, Alison, Ashkhain, Joanna, Peter, Rosemary. When we zip around the garden with the mower dodging this collection of flowering things, anyone else will think of them as simply plants, but I will know.

Glow worms

This night walk has a purpose; Georgie and Michael and I are out to find glow worms. We saw only one last summer and have high hopes for tonight. It is a night with a slow, quiet dreamy quality, a half moon, two cats twisting about our ankles as we leave the house.

We have torches, but there is no need to use them – we know the pathways so well. The shadows cast by the moonlight distort pebbles in the path, throwing thin black streaks off the grasses, and the cats dart past legs and through fronds like little streaking devils, making us jump aside, primitive beings spooked by the dark.

Where the path splits into three, we link hands and twist our way through jungly ferns. An unseen briar snags on my leg and I grope for a stick to clear my way. We are whispering, unwilling to pierce the magic of the darkness and the moon shadows. We edge our way along the upper bank of the river on a less-used path, which is dense in leaf and stalk and tugs on our ankles, and we creep forward slowly scanning, staring, prying into the dark to find a hidden glimmer of green, a pinprick of light.

And there it is. A tiny point of green is glowing on a stalk and we part the leaves to get close and see the little insect burning bright in love or lust. We sigh in happiness. We move on, to make a loop of the wood, and the last stretch yields four more glow worms in their silent eerie search for a mate.

A moonlit night of glow-worm love.

Bats in the cottage roof

The very lovely family in Owl cottage has approached tactfully to tell me that we have rats in the roof; they can hear them scratching in the ceiling. I am quick to reassure them that it is bats that they hear, not rats, but the idea of bats seems to instill horror. I think they imagine vampires.

I tell them that bats are a protected species and we are so very lucky to have them on the farm, but that I will ring the Bat Protection Society to get advice.

The holiday family look chastened but relieved and I slump off thinking that I have a million more pressing things to do. The Bat Society grasps the problem in an instant and say they will send someone round at dusk to count out the bats and then seal up the entrance hole through which I have seen them coming and going for several years. Apparently, they will simply relocate.

Michael is trying to hustle me out for a date night, so when the bat man arrives, I am a bit hassled; I take him out to the gable end of Owl Cottage and point to the offending hole. He has a ladder and nips up to do his stealthy work and tells me he will be back tomorrow to check that no bat is still left inside. He also tells me how lucky I am that it is not springtime, when there might be a bat nursery full of bat babies in there and he would not be able to get them out.

We leave him to his work and the next evening when he returns, he says they have all gone from the roof. In the twilight we can see bats flying and we linger to see where they have chosen as their new home. They have simply moved over above the other bedroom, through an almost identical hole on the other side of the roof. We cross our fingers and hope that the children in the bat bedroom are better sleepers than their parents.

Clearing the orchard

It's a bonfire day. We will clear and burn the trees in the old orchard to make way for a new drive to the front of the house. It's a huge project, planned, discussed and mulled over, and today we are dressed accordingly. We have lined up a team of helpers in the form of kind, loyal, and energetic friends and fit young men, who look more keen on the daughters than the project.

There is a considerable pile of grubbed-up tree roots, all that remains of the old fruit orchard. We had hacked at the diseased boughs and pulled up the stumps and dragged the earth-clogged roots to this bonfire pile. It was a massive job. Colin had come down with his tractor and threaded a chain round the stumps and revved and pulled.

They were deeply reluctant, those deep roots, and Colin's tractor had struggled and bucked, making deep furrows in the soil. Slowly and with a moan, the roots had yielded. The field was scattered with the fallen and we had dragged them to this haphazard pile.

The tough jobs are to take a bow saw and reduce a tree to logs, while the lighter jobs are with loppers and secateurs and wheelbarrows. I see Michael astride a tree that is balanced on a tumble of roots and he is wielding a saw; Will is tackling a lower branch and their counterpoint seesawing is terrifying. I scurry about with the loppers, diving into the tangle and attacking branches that are hampering the serious work.

Apple logs are beautiful and give off a fragrance of sappy sweetness. They will be fabulous on the fire next winter. We are sorting the small bits into heaps for kindling. The scrappy little twigs will form the heart of the bonfire, and we need to decide on a site for it. The huge pile of stumps and roots will gradually be shifted as pieces are cut and stored.

The bonfire starts as a scraggy pyramid of knobby fingers, twigs, and shrivelled fruit buds. To get the flames going, there are bunched-up pages of newspaper, broken shards of a wooden pallet, and a fierce competition to be the one to light it. Matches won't light in the wind, and a lighter has singed the hair on Will's hand, but laboratory tapers are looking promising. The newsprint flares up and we think it is lit, but no, it is snuffed out by the wind. We huddle round to make a windbreak and poke a hole into the heart of the pyramid of twigs, and then thrust in a cheating barbecue firelighter.

Whoosh! Primitive stuff, this fire-making. We rush to get extra logs to burn, needing a huge blaze as a reward, to connect us to our deep past. A mouldy hay bale is broken into tranches and packed onto the sides to focus the flames on the tightly packed twig heart.

Our fire is roaring, it's tempting to forget the lopping and sawing and simply hurl great branches onto the fire. We lurch from one task to the other as the day lengthens. I bring out big mugs of tea and we stand around, our faces gold from the setting sun and the loom of the fire. This bonfire will be burning still in the morning and the job can begin again.

Advice from friends

Usually I love our friends, that loyal team who heave-ho down to Cornwall, laden with hugs, smiles, and exotic foods, vaguely willing to share farmyard toil, but longing for country air and a rest. But today I hate them all.

We have a gathering of friends, with much larking about, but at dinner last night they got out of hand. I had shown them the architect's drawings.

These are the blueprint for our future here, for our old age, for an easier life. We could sell the farmhouse, the small paddock, half of the big meadow and the triangle orchard, and keep the drive, the new field, the vegetable garden and the remaining half of the meadow, as well as the cottage courtyard with three rentable cottages and the big barn, exquisitely converted into my dream home. The plans are fabulous. A vast oak door slides open onto the view along the valley; it's open-plan living the way we like it.

So why do our friends fuss and criticise? A torrent of objections flows forth. We are crazy, apparently. Having others living near will be a torment. The sale of the house won't balance the cost of the conversion. The accommodation will be too small. What's wrong with keeping everything as it is? Everything is wrong. The house needs a new roof that we can't afford, the sheer workload of running the cottages and farm is too onerous, and we want to let the cottages on long-term rentals; we are ready for a change.

I listen to eight people taking charge of my life and want to scream. Michael is so readily in agreement with the enemy that I suspect he never wanted this scheme in the first place. I hold my head in my hands, fighting back tears of despair.

I have spent hours on this project, gazing at the drawings, adjusting things with the architect, meeting the planning officer, talking with building controls, dreaming, dreaming. Dreaming of a quiet life with the long swoop of the valley in its shades of greens and purples stretching out in front of us, the vegetable garden, the copses, the flat part of the field ready for wedding marquees, the lawns and hedges waiting for grandchildren to climb over and make the same old dens.

I feel ridiculous and this plan is shredded in one night. We must sell it all.

Vegetables under the moon

My tyres scrunch on the drive; Woo has been waiting and trots up to rub her head on my leg. I've had a long old day at school, but I'm full of energy. I leave my bags in the car and skip down the steps and snatch a slice of bread and Marmite, before stripping off my clothes and putting on shorts and a T-shirt. Out into the garden. I lie flat in the grass and wriggle my toes. I bury my nose down to see what insect life I am disturbing. The sun is on my back and so too is Woo, curled up on me, a contented cat.

Michael arrives home and I look at my watch. 5 p.m. We have five hours of daylight to spend in summer occupations and I make him some tea to drink out on the warm slate bench, while we make plans for our evening. We walk hand in hand down the drive and seize a trug and some hoes and a kneeling cushion. We settle down close to each other by the beans to pick and weed and chat about the day. Murmuring about school scandals or strategies, we enjoy the sun and the whole, long promise of a summer late afternoon. Two cats are pouncing on each other and hiding in the rows of lettuce, and we are edging along the bean towers with piles of discarded weeds to one side.

I get up and stretch my back and go to fetch clean bowls for picking into. Michael wants to mow the vegetable garden, but I am persuading him not to, I don't want the noise or the scattering of grass bits over my vegetables.

We eat chicken and salad and a bowl of raspberries and cream out in the conservatory, then go back into the garden to savour a rising moon, huge and low. There are hours of daylight still. We decide to walk the bounds and then put the animals away. The evening is fragrant now and our hearts light. A few midges, our companionable cats, and vegetables under the moon.

Robert's dog

I zoom into the car and make a speedy trip down the drive and up the next drive into our neighbour Robert's farm. I have an urgent something to ask. I don't come here very often and I wonder where to stop the car. Dogs are barking from every direction and as I open the door, one comes hurtling towards me, so I shut the door quickly to consider my next move.

I think the dog might have gone and I can see the farm back door ahead, so I reckon it's okay to get out. I have made three paces in the right direction when the barking dog rages close with its teeth bared, and a low growl to menace me. I get all brisk and tell it I am here to see Robert, hoping it can't tell I am terrified. Three more paces and it runs at me, sinks its teeth into my leg and won't let go.

If I were a rat, it would be shaking me. I am in shock, in pain, and I see Robert at the door. 'Call your dog off, call it off!' I shriek, and it seems that he takes forever to yell at the dog, whistle loudly, and then it lets go of me.

'For God's sake,' I shout, 'it's bitten me!' I can see the blood welling through my torn jeans.

Robert is silent for a moment and then tells me, 'You should have put your face down to he, that would have stopped he.'

Trembling, I get into the car and drive back home, fast.

I fall onto the sofa in the conservatory and yank down my jeans to look at my ripped leg. Do I need tetanus injections? Rabies? Calm down. My jeans are badly torn. There are puncture marks in my leg. I'm pretty shaky, and waddle into the kitchen with my jeans round my feet to make a cup of tea.

What did he say? Put my face down to a crazed, barking teeth-bared dog?

Am I that ugly?

Northern lights

It's too exciting: the *Nine O'Clock News* has revealed that it is possible to see the Northern Lights even as far west as Cornwall. We gather up a camping mattress and some cushions, garb ourselves in a rug and hats, and make a comfortable lying-down spot in front of the house, where we reckon we have the best chance of seeing it.

It's lovely to rest there, companionably, comfortable, scanning the sky. It's a good clear night, just getting beyond twilight, and there are stars. I think we don't mind if we do or don't see the aurora borealis – the stars and a mattress are quite lovely enough, but after a while we do sense a change in colour in the sky. It seems to be shimmering a greenish colour, it's faint, but definitely green and it is pulsing a little. After a while of green, the sky pinks up a bit and then we are back to what is sky colour. What is sky colour? We are momentarily confused.

Was that it, did we see the Northern Lights?

Tree pruning with Ezekiel

Michael is on a dangerous mission. The phone line keeps cutting out and when we trace the problem, we see that the branches of a tree are tugging at the cable. He drags out the longest of the ladders, the extending one, and wedges it hard into the mud at the bottom of the tree. He checks that it feels secure and instructs me to stand at the bottom rungs and hold it steady, then he shins up carrying a bow saw. I am to direct him to the right branches in case he can't see adequately from his lofty position.

I'm a bit worried: we are on a slope and he is very high up. He starts sawing and throws down a clutter of branches, some of which hit me, but most just fall to the ground. At this point, Michael's cat Ezekiel shins up the next tree and looks across at him and starts a distracting series of calls and chat. Michael laughs and gets on with the sawing and E climbs down. I think he has gone to do something more interesting, when suddenly he lunges up onto the rungs of the ladder and starts to climb up.

When he gets to Michael's feet, he squeezes round and upwards and Michael suddenly finds E by his face, beaming and purring.

'Hey, E, get down, Rosanne get him down . . .' But of course, I can do nothing.

Michael hooks the bow saw onto a branch, gives E a pat and a tickle, and moves him onto the tree trunk, but then E gives a merry leap back to the other tree, where he remains, shouting encouragement.

Snow, exhausted

Just lying in bed, I can tell that there is snow outside; a wonderful, heavy quiet, the world in muffled breath, a whiteness that seeps in under the curtains. But this morning I am a sluggard, so tired from teaching this past germ-laden term, and Morwenna home last night like a ghost. I heave myself up and see a breathtaking white world, the trees on the opposite side of the valley are white shadows, the river a grey pencil line.

With a shiver of pleasure, I realise that I don't have to drive to school, we will surely be snowed in and I have a day of snow play. I wonder whether to do the sensible and best thing and slip back into bed, but a day of snow is a once every ten years sort of treat, and it cannot be wasted in bed.

Michael can be heard in the yard, letting out any creature who wants to play in the snow, and I see a beautiful flight of white ducks running and flying down to the river, snow on snow.

I creep into Morwenna's room. 'It's snowing,' I say, very quietly, to test her. She opens a heavy eyelid and breathes out, 'Oh, how wonderful, let's go and play.' Georgie says the same, and soon we are sitting silently in the kitchen swathed in warm dressing gowns, eating porridge. We are older now. Only a few years ago, we would all have been out throwing snow well before needing breakfast. Older and much, much more tired. We all know we want to play, we are simply getting ourselves alive enough to do so.

Michael comes in with pink cheeks and laughs at us. He says he is walking to school and starts to select suitable clothing, a mix of smartish jacket and thick socks and wellingtons. I wonder how many pupils he thinks will show up. We try to persuade him that he is snowed in, but his sense of duty is too great.

As soon as we have our noses out of the door, we are alive. We pad around the edge of the house making heffalump footprints and then slide and scrunch through the snow to the big barn to find the sledge, and to make more temporary sledges out of plastic feed-sacks filled with hay.

Up on the top ridge of the field, the swoop of whiteness and the levelling-out flatness before the river look pristine, and oh, so inviting. Who goes first? We clamber on the sledges and roar down the hill, on and on, flying over bumps and ridges, throwing up a spray of snow in our wake, shouting with pure joy, and come to a gradual slow stop. I lie half on the hay-filled sack, half sprawled in the snow and stay there, with my heart racing and my mouth full of snow. Morwenna is halfway up the hill, eager for the next run and Georgie is close behind.

We have the whole day of snow ahead of us.

Cutting down the ash tree

I am so sorry. I am so very sorry. Ash tree of great beauty, we have to cut you down. Your huge trunk is bulging out from the bank and if we let you grow more, you will pull the bank down. When we first moved here, you were probably a little whisper of a sapling and maybe I should have taken you to a better place, but you can't stay here now. I am so sorry.

Colin can come and work on you with his chainsaw. We will have logs and more light, but I will miss you with your shadows dangling on the drive.

Colin arrives and shins up the trunk. His saw snarrs in the branches and one by one they fall to the ground. I am tense and send thought vibes to Colin to be careful and safe. More branches. More branches, like rainfall. When the small branches are stripped away, there is only a firm grey torso left, and Colin descends and starts to cut across the trunk at the level of the bank.

There is a huge roar, the primeval sound when a giant of a tree falls, our giant, our bank, and the ground shudders as it receives this great bulk. Colin is safe on the ground and we start to pull away the branches across the top of the bank, letting them drop into the log zone.

It doesn't look right. Protruding from the bank is the massive swelling of the base of the ash tree and it has a sharp vertical cut. We all stand back and look and I ask Colin if he could cut it flush with the bank. Suddenly Colin changes from farmer to fine sculptor and his saw sweeps through the trunk in a beautiful curve.

Three feet high and elegance itself, the ash trunk as sculpture stands at the join of the drive, and it becomes guardian of the house.

Holy Boy

I don't know how I find myself as Principal Shepherd in a nativity play. This should only be kids' stuff.

Thinking about it makes me feel sweaty and panicky. I don't mind the crawling about on all fours as part of the flock around the crib, bleating. I actually like that, but it's that this is a musical and I learn with a sinking heart that the play opens with me singing a faltering solo, 'Shepherd, Lonely Shepherd.' It's three whole verses long, all me, on my miserable wobbly own.

The first night in the local parish hall had piano and spotlights, with me swathed in rustic garb and my family grinning in the second row, and my reedy solo came and went. Then at St Cleer village hall, things went a bit weird. We had no piano but a portable keyboard, and in transport or perhaps by mischievous intervention, it had been set not to the piano sound, but music from the bottom of the sea.

The lovely lady pianist gave a look of horror as the opening bars gurgled and boomed. She didn't stop and say, 'Excuse us,' but gave me a firm nod that meant, off you go.

So I warbled 'Shepherd, Lonely Shepherd' in three verses to deep-sea noises. The whole of the first act continued to the sounds of the bottom of the sea, with Mary and Joseph, archangels and innkeepers all singing their hearts out to the burbling accompaniment.

The cast was writhing with suppressed laughter, trying to gauge the response of the crammed audience, when the room was suddenly plunged into blackness. No more bottom of the sea. We whispered urgent instructions to the technical team with no luck, until a booming voice at the back of the hall shouted out, 'Try a fifty pence in the meter!' Yep.

At half-time the crowd surged forward for tea and mince pies and we begged the piano lady to reset the keyboard. 'But I don't know how,' she said. We all gave it our best try for the duration of the interval, and we held our breath for the opening bars of the second half.

Bottom of the sea again.

Pasty lessons

Say paaasty or pasty? 'Paaasty' to be properly Cornish. I was sent to learn how to make this most essential food as Michael's bride, to spend the morning with Marion, his grandfather's gardener's wife, way down in the village of Praze-an-Beeble. She was the queen of pasty making, round and tight as a bun.

'This sort of pastry is mixed like this,' she said, 'cold fingers please, light touch. Chop your swede small, like this. The potato, too. Into a bowl for mixing with salt and pepper. The meat in fine shards, skirt only, it's the juiciest. Stir it all up. Roll out the pastry to the right thickness and turn it, turn it, keep the board floured. Cut round a small dinner plate, that's your pasty size folded over. Meat and potato and swede mix get put in the middle, none to go peeking out, fold it and crimp the edges with a smear of water and quick brisk fingers, fold, pinch, fold, pinch. Brush the surface with beaten egg, make a hole in the top, put the pasty on an old butter paper and in the oven it goes. Pop a knob of butter in the hole when you take it out. Mind you don't overcook it.'

Phew, I was whisked through the process, emerging red-cheeked, but bristling with pride and success and a tray of misshapen pasties. What could be better than to repeat the instruction for Georgie and her Nick on a cold, blustery visit?

'I've got skirt in the fridge, want to make paaaasties?'

Chapter 19

Estate agents

The estate agent's blurb for Large Bottom Farm of 1978 was dismal. It was a stapled square of pink card with a blurry photo of the farm. Inside it gave measurement of the hall and claimed fruitful gardens. Nowhere was there a hint of the magic of the place, the stillness, the sense that the house was quietly dreaming, the promise that our children could grow and play.

Now, thirty years later, we must play the estate agent's game again, with the noticeable difference that a glossy format is required and we have to pay for it.

The first agent is blank, efficient, soulless. He clearly doesn't like the place. The second agent is breezy and gushing. If we feel like this about the agents, what on earth will the selling process be like?

Eventually we warm to an agent who seems to like the house. This feels like a positive. The photographer comes round and lies on the floor and almost erects scaffolding to take deceptive pictures of the rooms. All the lights are put on to create glow. Obviously, the theme for selling the farm is that this is a house of grandeur, the wow factor is being fabricated in front of our eyes: vast rooms and baronial walls and endless romantic meadows. We roll our eyes and leave them to it. The glossy brochure falls on the doormat with a thud and our home is rendered unrecognisable, until we peer closely and see our kitchen table and the yellow sofa.

The steady trudge of prospective buyers then starts. We oblige by being tidy, making sure that either fresh bread has been baked or a cake, filling the house with tantalising homey smells, having charming countryside flowers cascading out of old milk churns, and making sure we are firmly out of the way.

We then get less enthusiastic and bother not so much with the alluring smells. I think back to the different house we ourselves viewed over thirty years ago, when on the third visit a small child was allowed to accompany us on our tour and showed us where all the damp patches had been newly covered.

And then, astonishingly, we have a real live offer of a spectacular sum of money for our house, the fields, the orchards, the kitchen garden, the barns and the holiday cottages. We accept.

We tell the girls. 'Oh no,' says Morwenna, disconsolate, 'Can I buy it?' No.

182 party

Hurrah, another party. We had grieved that the girls hadn't held enough outrageous debauches in the barn for all those years and now, just as we are poised to leave, Morwenna comes up with a spectacular plan. She has five London friends with almost simultaneous birthdays and we are to be the venue for a festival.

It is to be called 182, in honour of their combined ages or dates or something, and the big field is to be a campsite and . . .

Now the party has been and gone, there still seem to be assorted people hanging in there.

The barn has its Acrow prop supporting the floor above for late-night dancing, and is splendidly decorated with fairy lights and long threads of sparkly metallic strip. The stable part of the barn is sectioned off into a series of cosy sit-out spots. The power lines from the cottage laundry, both holiday cottages and the milking shed are a snake pit of electrical leads to DJ decks, to alarming attempts at spotlighting, and amplifiers and other things a live band needs.

I had dragged out various art installations (an outsize papier-mâché fist and the figure of the bus driver) to add to the vibe and people had hammered huge printed panels onto the goat-shed doors. Barbecques, little candles, sofas, hay bales to sit on, beer, wine, more beer, vodka shots.

And the drive quickly filled with cars and late-twenties cool professionals all as excited as kids. No rain. The squeamish and the princesses got beds in the cottages. We had mowed pathways to all the flat places in the field and tents sprouted like mushrooms. Food piled up, birthday cakes, sausages, slabs of doubtful-looking meat.

There wasn't the right electrical lead to get the live band connected. Phone calls, controlled panic, until one of Michael's ex-students came up with the right lead from his own band. Still no rain.

Wine and beer. Loud pulsing dance music. The neighbours had been warned. Friends were reunited, and everywhere hugs and love and dancing, all of us dancing.

Alone with blackberries

We are alone and there is a warm autumn sun and I am feeling happy. Michael has run out of jobs for himself and for me, so I suggest we go blackberry picking together. Secretly, I am imagining that this is the sort of thing we could do together when we have sold the farm and we are living in companionable harmony. We pick, silently, and the sun is warm on my back and there are butterflies close by.

Suddenly Michael gives out an agonized cry. What?

'I hate picking,' he says.

Wintery night

I saw a shooting star last night. I took it as a sign that things are going to get better. I stomped up the hill breathing in great gusts of frosty air, feeling the air cleaning and stretching my lungs. It was so dark I couldn't see the path, but I knew the way. I like that – knowing this space so well I could walk it in pitch-black, dead of night. Useful if avoiding a psycho killer out to get me. I would lurk in the nettles, powerful in my confidence.

I stared hard at the ground for a while to see if I could make out the shape of Woo following me darkly into the night; when I left, she had been slugged out on the sofa in a feline trance.

I gazed up at the sky, trying to identify more than my old favourites: the Pleiades, Orion, Ursa Major, Betelgeuse, the Milky Way. I saw two satellites flicking and twitching up there and a plane. In the half-distance, I could hear the roar of a train lunging past on the Liskeard viaduct.

Trains and starlight and grass smells is my habitat. I gazed at an ash tree lit up from below, throwing its leaves into a fountain of light beams. The apple tree beside it has three white apples, high up and gleaming like beacons. Bird food. I felt glad to be out there, longing to stay out all night, breathing in night air, night sounds, night beauty, until the time-to-go-to-bed voice took over and battled with my longing to linger, hugging on to the wild night that drifts and makes me whole again.

The morning is shimmering frost and birdsong sunrise. As I cross the Tamar, I look over the huge water hidden by rising mists and I long for a day of wandering on moorland and woods. I'm a caged creature. I'm the green man let loose in the classroom. The ivies and tendrils will break out some day and I'll roar off, wind-torn and into the woods, invisible.

Party games

I can go misty-eyed about party games; I have even read a small book with more extreme suggestions than those bubbling in my mind: it actually suggests that adults should shed clothes at the entrance and crawl on all fours behind each other to trigger companionship! Oops! Really?

I think of my game for six-year-olds, where I chose a different coloured ball of wool for each child and tied one end to a piece of furniture and tangled and wove them in a spider's web for them to untangle and found them in hysterics and utterly tied up half an hour later.

I recall the New Year's Eve parties for teenagers and doubtful parents, where we played Cracker Whacker. Sellotape a cracker biscuit to your arm and set out as a warrior with a rolled-up newspaper, hell-bent on crushing everyone's cracker but keeping yours intact.

Then there was Spinning the Trencher for those same teenagers, where you twiddle the circular breadboard and shout someone's name; if they don't catch it before it falls then there is the forfeit of a garment. What cheating, what cheating. Adults stay in the next room with wine.

My own father had an outrageous team game of dropping a golf ball into a bucket with a fine resounding clang, once it had been carried firmly gripped in the buttocks across the length of the room. The social standing of our family may have changed forever after we were asked to bring a game to a party and chose this one. Everyone else politely brought pencil and paper games.

So I was savouring what game we might play to enliven an impromptu party I have arranged for our very diverse group of neighbours: vicar, farmer, abandoned wife, horticulturalist. The adverb game seems okay. I write a collection of adverbs to put in a bag: *insanely*, *boringly*, *suspiciously*, *wickedly*, *wantonly*, and more. I suppose I might have been a bit carried away.

The neighbours tolerated the assorted food and drink and were willing to play games to humour their hostess, but the results were chaos. The vicar lay on the floor for *pointless*, the camellia lady writhed on the sofa for *saucy*, the farmer was accused of *stupidly* or *criminally*.

Michael and I wonder if our neighbours will speak to us again.

Last bonfire

We are assembled: parents and two girls, two boyfriends. There is work to be done and much romping and larking to do. They are like puppies released into the wild, buttoned into ramshackle jackets and hats, and off they go leaping in the field towards the mud slide. They squidge about in the marshes by the river and alarm the ducks that are hiding there. I am ordered to take photos of them clustered in front of the oak that is lord of the bridge and then we all slither our way along the cattle-trod path. We reach the mud slide, where the rope swing no longer ends in a T-bar, but we jump and hang.

The big project of the afternoon is the last bonfire. It's a therapeutic sort of bonfire, burning the food barn's last dusty, mouldering rubbish, keeping the fire big and high with laurel branches or bamboo canes, the Rosebud wooden sledge full of woodworm, old string and wisps or tranches of dusty hay, and old cardboard boxes. Burn, burn.

We are clustered around, watching in satisfaction as the light falls and our faces are glowing. Great spirals of red sparks rise up into the sycamores above and scatter ashes. We come and go, bringing more fuel. We each have a stick to poke errant burning things from falling out of the fire. We group and regroup, calling out if anyone has found a new source of stuff to burn, old dried stems of maize, anything. Exhilarating, compelling. The moon rises huge over the valley.

Bat story

I am marching down the drive clutching a Dyson vacuum cleaner and my thoughts are on murder. Bats, you know I love you, and I know you do not live in the goat shed, but watch out, because I am coming to swoop on the inside of the goat-shed roof and will not be held responsible for my actions.

At this very moment I am caught, looking guilty, by Louise who has dropped by. 'You look armed for trouble,' she jokes, and I abandon my attack and traipse into the kitchen to tell her our woes.

The farm is sold pending signing of contract. The buyers are suddenly dubious about watertight planning consent on the goat shed and require fresh permission. Exasperated, we agree to reapply and back comes the news that this is now a new permission and as such needs a bat survey. Bat survey? I choke over this. We have no bats in the goat shed; we have bats everywhere else. We have bats in the cottage roof. We have bats in the farmhouse roof. We are friends with the Bat Society. We are protectors of bats and knowledgeable about bats and we definitely have none in the goat shed.

We still require a Bat Survey. A man who is expert in bat things will come and inspect and we will pay him a fee for this and his subsequent report, which will come in at less than a thousand pounds. I am apoplectic and powerless, it seems.

When he came, I escorted the bat man to the goat shed, explaining our great sympathy and love of bats. I suspect he was going to find evidence of bats even in the middle of an ocean.

The goat shed was as usual full of goats and hay and chickens, and a dusty manger with cobwebs on the rafters and bits of baler twine. He busied himself while I stood watching, leaning on the door frame. He bent down and looked carefully at the ledge of the manger, then the top of the manger, and he poked about and scuffed through some musty hay, getting on his knees to scrape away layers of ancient midden.

Then he stood up triumphantly holding a piece of card, on which lay the smallest and single oval of bat poo. Surely mouse poo?

'No, it's bat poo,' he said happily, scientifically. 'One bat poo?' I puzzled back at him, 'One bat poo?'

'It's all we need,' he said, 'it's evidence that bats inhabit this barn.'

Oh, no. I explained to him that of the hundreds and possibly thousands of bats who inhabit the area that is our farm, it is certain that at least one bat would have found it possible or attractive or necessary to fly through the building, but bats do not inhabit this barn. I'm certain, we are here. We know where the bats fly. This must be a fly-through poo.

The bat man had given me a superior look and said the results of his inspection would be sent to me and to the planning office in a report.

I told all this to Louise, with my head in my hands. I had spoken to Natural England on the phone to hear what the whole process looked like and had learnt that after a report which found bats, we would need two audio recordings at six-monthly intervals to check their density.

'The point is,' I said, 'it's likely to wipe out our farm sale.' Louise sighs. I sigh. We drink tea. 'What were you going to do with the vacuum cleaner?' she asks.

Clean up the barn, but of course, it's too late. It was my gesture of rage.

Ha! Bat story completed

I have moaned and got serious and highbrow and full of bat wisdom and persuaded the Higher Official at the Bat Office that the six-month interval between audio tests for bat flight in the goat shed was quite unnecessary and the tests could be run concurrently, a solid six-week stretch of bat recording.

They seemed unsure, citing the fact that the bats would not be flying so much in the winter months. I returned a volley, pointing out that a single bat poo, possibly lying in the goat shed for a period of several years, did not constitute much of a problem. So they agreed and the bat man returned to fix a drab khaki box onto the top of the manger with some duct tape.

Six weeks. He came and took it away yesterday and told me the results would be known shortly.

There he is now, on the phone, and his news is that there was no flight of any bat recorded in the six weeks. Told you.

I wonder if being a bat inspector would be a lucrative job for Michael in his retirement.

Leaving ducks

We need to find new homes for all the critters. They must go because we must go. We have shed creatures before, when there have been too many or someone has wanted to buy; we have culled almost everything if we needed meat or when old age overtook a lingering animal, and we have even dispatched cockerels as severe punishment for bad behaviour. But we feel tender about our assorted flocks and herds, and today we are bracing ourselves to hustle twelve ducks into the van and deliver them to a good home at Lantallack.

This is absurdly simple, the trail of food is an easily followed pathway to the van, the future surely is bright for a well-behaved duck. But we have heavy hearts, our innocent ducks will not know the new dangers.

Letting go is harder than I thought.

Last wintery days in empty house

We are in a Hitchcock movie, definitely *film noir*. Michael and I are sitting in the drawing room on a deeply wintery day and there is a fire smouldering in the grate, but it has no fiery lust. We each sit on a kitchen chair, I am wearing Granny's full-length mink coat made of something that isn't mink, collar turned up high. In an attempt to keep in a little warmth, I have tied the bottom of the coat around my ankles with baler twine. Even Michael, that impregnable fortress, is cold.

Almost all the furniture has gone; we are sitting on the two chairs that we are going to give away and we still have the television and VHS system, which we think will moulder in storage. There is a mattress, some bedding, and some meagre provisions for breakfast.

A long year of severance is over. We have taken barometers, good paintings, my grandmother's priceless sofa, and assorted brown furniture to grand salerooms, and had to bring back the round table and the priceless sofa. We have unloaded countless collections at the humble local saleroom. We have lovingly donated farm implements on family and friends. We have sold so much stuff in surprising directions, including the Cricket Field car-boot sale, where Richard and Judy and Michael and I appalled the other stall holders by selling every item for 20p. We were regular customers at the recycling centre, the charity shops and the dump.

We have had many bonfires. Morwenna's anatomy notes have gone up in flames and so too have Michael's and my past careers. We have felt purged and renewed. But now on this wintry day, we are feeling cold. Tomorrow we leave our home, home for over thirty years, home where we raised our two little girls and countless critters, home where we have tilled the soil and wandered free.

Tomorrow we will think of our daunting commitment to volunteer in Africa, but tonight we reach out and hold hands, our icy hands.

Acknowledgements

Thanks to my friend Amelia Fletcher of the sharp wit for reading and laughing, and to all our friends who came to Large Bottom Farm and worked and played on the land.

Thanks too to my delightful agent Florence Rees for believing in me and to Hannah and the team at Hodder for welcoming me into the fold.

Thank you so much to our wider community around Liskeard for grounding us for over 30 wonderful years. So much of what is written was quite long ago but is as fresh in my memory as if it was yesterday.

And a loving thanks to my long-suffering husband Michael and our beautiful Cornish girls Morwenna and George.